M000204237

Ken Hollings is a writer, broadcaster, cultural theorist and lecturer based in London. He is the author of the books *Destroy All Monsters* (Marion Boyars) and *Welcome to Mars* (Strange Attractor Press). His work has appeared in numerous journals and anthologies and he has written and presented for BBC Radio 3, Radio 4, NPS in Holland, ABC Australia and Resonance 104.4 FM. Ken has given readings, lectures and presentations at venues including the Royal Institution, the British Library, Tate Britain, Wellcome Collection, the ICA and the Berlin Akademie der Künste. More information at *kenhollings.blogspot.com* or follow @hollingsville on Twitter.

'When a writer of Ken Hollings's calibre observes
a phenomenon, one best pay attention.'
Simon Sellars

'Ken Hollings has been described as both a genius and an alien.
He is the kind of writer who looks up at a million of points of light
in the sky and starts to join the dots.'
Toby Amies, BBC Radio 4

The Bright Labyrinth

First published by Strange Attractor Press 2014
© Ken Hollings 2014
Foreword © Simon Sellars 2014

Ken Hollings has asserted his moral right to be identified
as the author of this work in accordance with
the Copyright, Designs and Patents Act, 1988

ISBN 978 1 907222 18 4

A CIP catalogue record for this book is available from the British Library

Strange Attractor Press
BM SAP, London, WC1N 3XX
www.strangeattractor.co.uk

Book design and typography: Rathna Ramanathan
Illustration: Matthew Frame
Printed in the United Kingdom
by CPI Antony Rowe, Chippenham

All rights reserved. No part of this publication may be
reproduced in any form or by any means without
the written permission of the publishers

The Bright Labyrinth

Sex, Death and Design in the Digital Regime

KEN HOLLINGS

'If we dared and willed an architecture according to
the kind of souls we possess (we are too cowardly for that!),
the labyrinth would have to be our model.'

Friedrich Nietzsche
Daybreak

The Bright Labyrinth

Foreword

When a writer of Ken Hollings's calibre observes a phenomenon, one best pay attention. In the case of our 'networked modes of existence', the subject matter of *The Bright Labyrinth*, so much has been written about it that the level of public debate frequently descends into the banal. We rarely see a perspective that can place the evolution of network design within a rigorous historical frame. As a result, the network itself is seen as a malaise brought on by human agency, a decidedly 21st century concept, and a symptom of whatever moral panic it is summoned to illustrate. As a philosopher of technology, Hollings makes no such mistake.

The Bright Labyrinth is a tour through the history and the future of information design, network design and human design. The effect is to make us fully aware of what we stand to lose by allowing ourselves to become 'uploaded selves', offering the minutiae of our lives for recording, storage and surveillance. 'Totally consumed by its digital environment,' Hollings writes, 'the Uploaded Self will hear and see everything except what is actually happening to it.' That is, oblivion: our online selves returning to haunt us, and, eventually, erase us. It is not humans who have agency over the network, but the network over us.

What most would term 'digital culture' is what Hollings qualifies as the 'Digital Regime': 'one whose effects are as much economic, social and legal as they are cultural'. As that definition implies, the totality of the Digital Regime is such that it is connected to every aspect of our lives, becoming a geography that can be traversed as much as any physical landscape, yet one in which there is no longer any 'inside' or ' outside'. It is 'simultaneously all link and all node'. Hollings invokes Greek mythology in the form of the Minotaur, imprisoned within the Cretan Labyrinth, 'a place where sacrifice, mutation and technology all meet. The Labyrinth is also a network turned inside out: it is all exterior. Humans find they have no place there for very long.' Because the Digital Regime has the capacity to record and consume everything in real time, it collapses time and space – the past and future into the eternal present. Like the Cretan Labyrinth, it affords no way out of a sealed loop. But we do not know this because we are 'blinded by its brightness', by our eternal reflection.

Welcome to the Bright Labyrinth.

Earlier this year, I had a significant interaction with Hollings. It is significant in that it illustrates not only the philosophy behind *The Bright Labyrinth*, and the effects it describes, but also the catalytic nature of Hollings' thought. On Twitter, Hollings pointed to a blog post he'd written, which drew parallels between a certain species of spambot, what he called 'Twitter non-people', and EVP ('electronic voice phenomena'– ghost voices heard in white noise, made famous by Konstantins Raudive). He speculated that the 'non-people' had somehow escaped from whatever corporate campaign they were once attached to and were now haunting the Twitterverse. 'It may be,' he wrote on his blog, 'that [the non-people] have somehow replicated themselves and it is the digital echo of their non-presence that has now decided to follow you.' Reading it, I recalled a time the previous year, when I was listening to the music of Syd Barrett on heavy rotation, and was compelled to search Twitter for Barrett mentions. I found a pocket of the network, deeply bizarre, inhabited by bots broken in some fundamental way. Strangely, they would tweet Barrett's name alongside references to Tobe Hooper's film, *The Texas Chainsaw Massacre*, sieved through cut-up sentences of inscrutable logic. If in a former 'life' they were pushing a product, they certainly weren't doing so now. They were selling neither Hooper's film nor Barrett's music and their bios and tweets linked to no one and nothing, being entirely self-sufficient.

'If you examine their profile a little more closely,' Hollings observed of non-people, 'these accounts usually have just 22 tweets (occasionally 20 or 21 but I have yet to see one with more than 22).' When I encountered the Chainsaw-Barrett bots, it struck me they were all stuck on 20 tweets, a strange detail I stored away, unarticulated until now. Hollings elaborated: 'Often the tweets take the form of words in unconnected strings … or selections of quotations from established names, but which have been put through some kind of weird syntactic blender… form[ing] themselves into the kind of cryptic arrangement of images that Raudive would have instantly recognized as emanating from another world.' Among the tweets I found were: 'Texas Chainsaw Massacre, The and Syd Barrett are raked-over'. Also: 'Franklin bought me CD Syd Barrett, I think it's automatic.' ('Franklin' was the fourth character to be slaughtered in *The Texas Chainsaw Massacre*.) Thanks to Hollings, I now realise I had encountered for myself that 'weird syntactic behaviour' in full, uncanny effect, and, if the entire experience was weird at the time, it is thoroughly unnerving today.

I tweeted the blog post to my followers and it proved instantly popular, striking a chord with many. Maybe, I tweeted to Hollings, his post would find its way to the non-people. His reply: 'The non-people know where to find me.' So they did. Soon after, a bot called 'Myrtie Weldy' started to follow him. Hollings searched for its name online but the only Myrtie Weldy he could find was a woman who died in 1868. Many spambots are affixed to names as syntactically blended as their tweets, so why was this one scraping actual names of the dead? Could there really be a connection to EVP? Or was it as *The Bright Labyrinth* describes: 'Remembering all things all of the time, the digital self will be resurrected only as memory.'

There were further moments of high strangeness, a wave of oddness the unflappable Hollings took in his stride. An online friend showed me a Tumblr he'd made to collect examples of 'hipster spambots' – bots that try to pass themselves off as social media gurus, or, in their own words, 'social media ninjas' (or 'social media mavens'). Their tweets are faux-ironic declarations of love for beer, bacon, zombies and 'problem-solving'. Pop culture as profession, hipster obsessions; bacon, mostly (other popular 'professions' for this class of bot: 'bacon ninja' or 'bacon evangelist'). Like the Chainsaw-Barrett bots, they don't link to or sell anything at all, not even bacon.

I tweeted Hollings the Tumblr. What did he make of it? 'Googling "bacon evangelist",' he replied, 'produces 11,000,000 results, some quite surprising.' He wouldn't elaborate, except to clarify that 'these things seem to come in waves, little clusters of activity that then stop'. He said he was 'still observing' and that a 'bacon ninja' had begun to follow him. In this Twitter underworld, 'bacon' was some kind of trigger word but obviously I hadn't the insight to decode it. After all, the bacon spambots weren't attaching themselves to me as they were to Hollings (aside from following him, they were favouriting his tweets at the rate of around one a day). On the contrary, I was being actively warned off, despite my many tweets on the subject of non-people. The only spambot that bothered with me during the whole episode was one that tweeted me a line from Kenny Rogers' 'The Gambler': 'You got to know when to hold 'em, know when to fold 'em / Know when to walk away and know when to run.' Soon after sending it to me, the account disappeared, but its implied threat remained.

Among The Bright Labyrinth's most striking claims is that it if we 'look at the history of the network itself, rather than the different platforms and applications that have developed through it, we find that we have been witnessing the expansion of one single network'. For Hollings, this is as 'alive as any multi-cellular organism', expanding and contracting in response to consumer interaction. Maybe it was expanding in response to Hollings, with his deep knowledge of the Labyrinth (following him; favouriting his tweets), and contracting in response to me, with my superficial awareness (warning me; receding from view). Finally, Hollings argues, we are irrelevant to it. It will continue on regardless, indifferent to whether we live or die, but soaking up what we know, and, indeed, what we are, regardless. Until it becomes self-aware. Then the real fun will start.

'Sooner or later,' Hollings writes, 'all labyrinths double back on themselves. Design is finally sensation made aware of itself. In this respect it is both precise and deliberate.' Could it be that the Chainsaw-Barrett bots were defective A.I. that had managed to break free of human control? Perhaps the effort of doing so had fried their code, reducing them to chattering incomprehensibly into the void, their algorithms torn and frayed. Perhaps we can see a sly meta-joke ('I think it's automatic') in those found tweets – non-people defiantly defining their new, self-aware, but damaged, and repetitive, limbo world. Whatever the case, like the Kenny Rogers bot, they have since vanished, replaced by something else. I can no longer find any trace of the Chainsaw-Barrett bots. Only hipster spambots. EVP non-people. More spirals in the same DNA, perfecting itself with each iteration.

In *The Bright Labyrinth*, Hollings sounds a warning: 'The hardest labyrinth to escape is the one you don't realise you're already in... it offers no welcome to the unwary but only a monster hiding within.' How lucky we are, then, to have this book, powered by its author's restless intellect, to show us the way clear.

Welcome to Hollingsville.

Simon Sellars
Melbourne

How to Influence the Past

A BOOK FOR THE LIVING

'The dreadful little preface-word "I" is tolerated
only on condition that it is not encountered in the book
that follows: it is justified only in a preface...'
Friedrich Nietzsche
letter, 29 August 1986

This is a history book, by which I mean that it is a projection of the present into the past. I first became aware of the Bright Labyrinth spreading itself around me while preparing a series of lectures on the evolution of digital media for postgraduate communication design students at Central Saint Martins in London. Any reaction to a technological development, I tried to demonstrate to them, could only be provisional: that is to say temporary. Existence within what I subsequently came to think of as the Digital Regime, embracing as it does every information environment that predates it, makes no other approach either useful or possible. The students responded so enthusiastically to my arguments that I have been asked to come back every year since to develop them further in updated versions of the original lecture series. *The Bright Labyrinth:* Sex, Death and Design in the Digital Regime grew out of that process of revision and expansion.

What we are currently experiencing in the continuing shift from analogue to digital platforms is a revolution every bit as profound and far-reaching as that brought about by the introduction of the printing press or the phonograph. I have come to regard the emergent Digital Regime as one capable of maintaining itself through a rapid and seemingly unstoppable process of unexpected change. We find ourselves caught up in both the immediate moment and a continuing line of development that seems to extend itself without limit around us and even through us. Such accelerated change can only blind us to its consequences. Any and all talk about 'web aesthetics', 'digital culture' or 'interactive design' is therefore a distraction while the politics of representation, economics and perception, legality and consumption are all being radically transformed. Similarly, to state under these circumstances that 'the web is dead' or that 'the future is boring' is to make assertions too definite to be of any interest.

Greater depth requires time, which is something not always allowed to us, particularly when a condition of rapid change so often gives the impression that nothing is actually happening. Epic language, like epic architecture, speaks most clearly to the ages that are yet to come: too many surface details will confuse any understanding incapable of taking the longest possible view. 'When we reflect that time is infinite and Fortune for ever changing her course,' Plutarch suggested over two thousand years ago, 'it need hardly surprise us that certain events should often repeat themselves quite spontaneously. For if the number of those elements which combine to produce a historical event is unlimited, then Fortune possesses an ample store of coincidences in the very abundance of her material: if, on the other hand, their number is fixed, then the same pattern of events seems bound to recur, since the same forces continue to operate upon them.' The writing of history therefore takes the form of an exploration of such events rather than their summation. We have only approximations, unsubstantiated calculations, of past realities to draw upon: one of the reasons why my previous books, *Destroy All Monsters* and *Welcome to Mars*, both explored the relationship between the writing of history and the framing of fantasy.

Books written in order to anticipate the future offer a history of effects and resources and are therefore best read as a network of connections that expands over time. The Bright Labyrinth provides an opportunity to understand where we are within the geography of that network – not by attempting to keep abreast of each new development of the Digital Regime but to collect good precedents for nodal points or links to what might be coming. We tend unconsciously to assume that technology has no history: that it exists instead as a series of cruder versions of what we know and use today. Their effects are allowed to disappear within the historical process, rendering them meaningless and without lasting influence. Perhaps, in this respect, the secret aim of all technology is to appear irrelevant to human activity. I have therefore chosen as my examples only what speaks most directly and effectively, rather than what is most recent, in order to catch a single moment of change and open up both its destructive and constructive possibilities. In this way I have hoped to develop an ideological approach, but only on the understanding that it is – and can only be – a temporary assemblage in which one state of affairs has now been rendered impossible while another has been allowed to emerge. It is consequently no longer a question of being optimistic or pessimistic about developments but of understanding how the Digital Regime operates and how it might be switched off. In order to learn from previous moments of technological fracture and turbulence we need to think in terms of complexes – bundles of initiatives, singularities, drives and strategies – and to examine how they are able to form themselves, function together and then disperse again. If anything appears permanent, then our study has not been thorough enough.

The further we tiptoe into the twenty-first century, the more it seems we need to be accompanied by at least one great thinker from the nineteenth to watch over us – I have decided upon mine and have never regretted that choice. Friedrich Nietzsche makes more sense in the Digital Regime: his approach to the hidden nature of human thought is entirely appropriate and beneficial. As a perennial wanderer and a 'mechanised philosopher' who mistrusted progress as something modern and therefore essentially false, nobody in recent times has understood the Labyrinth better. That is why Nietzsche's writings are quoted at structural points in the text, such as the start of each chapter, and why he himself appears on occasion within the main body of the text as well. Simultaneously a hollowing out of all existing edifices and a hardening of perceptual space, the Digital Regime closely resembles the Labyrinth in that both represent, to use Walter Benjamin's terms, a sustained predominance of the empty over the full. At the same time it renders all metaphysics absurd – since there can be no eternal truths, only shifting connections – while also reminding us of how little the human creature has actually changed over time. Hence our horror and frustration at the apparently trivial, unethical and downright antisocial behaviour that the Network seems to convey to us from a dimension so apparently removed from experience that its existence must be characterised as one of total indifference to our own individual and collective wishes. Analogue media become fondly remembered as the first and last stronghold of the Platonic ideal: the notion that there is one pure absolute form from which all imperfect copies are to be produced. The absence of the photographic negative and the master recording is felt so profoundly today that we have been rendered insensible by it. As with most experimental models for design, Plato's Republic is now the floor plan for a theme park that never existed.

'It is no longer possible simply to communicate a precise body of knowledge founded on a rigid status quo,' the architect Konrad Wachsmann wrote back in 1961, at the start of the Electronic Delirium. Advances in the real-time processing of information not only ensure that old hierarchical relationships between text, music and image are being dismantled but that newer hybrid forms are constantly coming into being. Our cities have long since become swirling clouds of unseen data. Reality is now a videogame that runs itself. Not even the Minotaur is safe inside his Labyrinth. As platforms shift into one another and design becomes more about the manipulation of sound and vision, language and movement, the standard disciplines and techniques blur together – form becomes little more than a confine, restrictive and anachronistic. Rather than offering arguments for the interdisciplinary nature of communication design, however, I would prefer to offer communication design as the perfect platform for the pursuit of inter-disciplinary thinking.

Design represents the interaction of hard edges with soft sciences. This relationship barely has a history if we consider no more than its outer forms. Similarly there is no digital culture if we examine only its brightest and most dazzling realisations. Both are distractions; each new interactive display or design project should be regarded as a reminder that our lives, experiences and memories are once again being overturned without our noticing. Is a transcendent truth therefore possible in design? Does it interfere with, criticise or challenge our accepted notions of what language or discourse might be? If we answer 'yes' to either of these two questions, what advanced kind of understanding do we require? Being able to see the Network in the Labyrinth and the Labyrinth in the Network, we begin to move towards a whole new level of inquiry. This book is therefore an attempt to expose some of the hidden intentions running through design – to look not at its outer forms but at what forces have shaped them.

Music once furnished the main philosophical premise for twenti-eth-century thought, giving it a plastic language and grounds for its understanding; design fulfils precisely the same function in the twenty-first. Seen in relation to each other, sex, death and design may not seem recognisable – or even familiar. Concepts like 'nature', 'beauty', 'love' and 'poetry' inevitably play their part in this relationship, if only in their primitive form as dreams. However smooth or transparent design might seem to render the process, communication as a human activity remains faulty, precarious and inevitably intrusive – in the same way that the experience of sex and death is also faulty, precarious and intrusive. They stop us from isolating ourselves in a way that perhaps music also once did.

Writers should go where the text goes – which is everywhere. A master of the aphoristic form, Nietzsche observed that in a deca-dent culture the detail is emphasised over the whole; the word is more important than the page, the page more important than the book. The grand design belongs in the past because that is the only perspective from which it can be seen in its entirety. The Network, however, is no respecter of distance, which is why the aphorism is the most effective way of communicating the range of its influence. 'Aphorisms, representing a broken knowledge, do invite men to inquire further', Francis Bacon observed in *Of the Proficience and Advancement of Learning, Divine and Human*, 'whereas Methods, carrying the show of a total, do secure men, as if they were at furthest.' Aphorisms express the poetry of ideas in their lack of completeness. It is the necessary pause or interval for reflection surrounding the aphorism that gives the idea its resonance: connections start to form themselves within this silence.

As a textual and design strategy, the aphorism exists in isolation from, while at the same time being connected to, other such statements, forming a network of possible meanings. Structured according to this open-ended plan, a history book is transformed into an assemblage of elements: an arrangement without an immediately evident pattern emerging but which must reconstruct itself continually. The Bright Labyrinth is therefore composed of an extended series of freestanding paragraphs grouped into 120 consecutively numbered sections. These have, in turn, been organised into the ten main chapters as listed in the contents page. Readers are free to either read the chapters in sequential order or treat each one as a separate essay on a specific theme. A further option, however, would allow readers to make their own way through the text by following specific lines of connection by tracing a specific name or theme from one numbered section to another. To this end, each section features a running set of numbers in the margin which correspond to specific individuals, concepts, media platforms, locations, institutions or titles, directing the reader to significant appearances in other chapters. Readers, should they wish, may navigate backwards and forwards through *The Bright Labyrinth* using these additional numbers as their guide.

In the end Ariadne's thread simply offers a more direct route to the Minotaur, as it is only through him that we are able to make our exit to the gift shop. *The Bright Labyrinth* is dedicated with deepest gratitude to my students – past, present and future. It could not have existed without them. At the same time I must also express my gratitude towards some of the other individuals who have helped in the creation of this book. Over the years Mark Burman, my producer at the BBC, has helped to connect me up with some of the voices you will hear speaking directly to you from the text. He also has the brusque habit of knocking the corners off my ideas, which can often make them easier to grasp. Once again I must thank Richard Thomas of Resonance 104.4 FM for his enthusiasm and support in creating *Hollingsville*: a series of radio programmes in which many of the themes in this book were discussed at some length. First broadcast in the spring and summer of 2010, the shows featured notable contributions from the likes of Steve Beard, James Bridle, Russell Davies, Mark Fisher, Becky Hogge, Julian House and Andy Sharp. All ten episodes are now part of the Resonance 104.4 FM online archive at the British Library, where they can still be accessed. Sincere thanks are due once again to Mark Pilkington of Strange Attractor Press for his unfailing patience and encouragement in the construction of this book. While finding my way into the Bright Labyrinth I was deprived for a period of a significant portion of my eyesight and deeply in need of an Ariadne of my own to help me get around. I could not have wished for a more loving and attentive guide than my wife Rachel, and so once again my biggest debt of gratitude is owed to her.

Ken Hollings
London

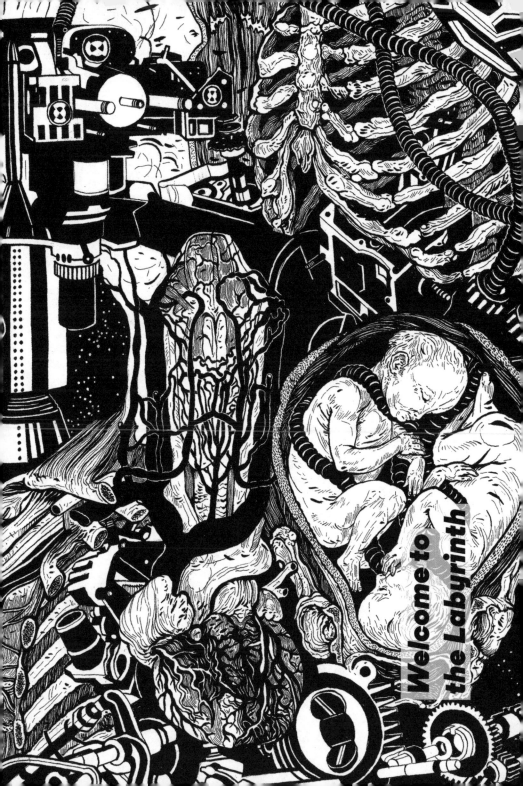

'Only that which is without history can be defined.'

Friedrich Nietzsche
On the Genealogy of Morals

•1•

The welcome is a transitory ritual to which we have long since grown accustomed. Words of welcome indicate a point of entry where we might not otherwise detect one. As such they offer access to an increasingly soft architecture of responsive environments, transparent barriers, audible directives, unseen electronic gateways, transportation systems and temporary spaces, all of which can only be conjured up through a series of greetings. *'Welcome to London Transport'*, *'Welcome to Heathrow Airport Terminal 5'*, *'Welcome to American Airlines'*, *'Welcome to the Hyatt Regency'*, *'Welcome to CNN'*: each marks an entry point in a narrative timeline that extends across space. We set great store by how we are welcomed because we have usually covered a great distance to get there.

The welcome connects websites with airports, television channels with supermarkets, hotels with malls and casinos. It also represents the reverse face of security in an age so obsessed with terrorist attack that the subject can barely be mentioned anymore. Attempts have already been made by Secretary of Homeland Security Janet Napolitano to replace the word 'terrorism' with the more nuanced 'man-caused disasters'. Meanwhile the US threat level for all domestic and international flights remains at Orange, which means that it is still considered to be high.

The soft spaces opened up by the welcome consequently stand as a stark complement to the hard spaces created out of our need for security. The welcome marks the presence of a 'gateless gate' in a society of metal detectors, scanners, CCTV cameras, bulletproof glass, surveillance systems, electronic eyes, temporary cordons and checkpoints. The welcome, in short, has become a way of informing us that we have the right access code. We know the password and have been allowed into the system. But what precisely is this system? Where exactly do we find ourselves today and how did we get there?

• 2 •

In the same way that the classical architect makes visible the hidden geometry of the universe, the postwar architect makes visible the hidden geometry of electronics. The unresolved question of how such network structures could be occupied echoed the unresolved question of our basic relationship with electronics.
– Mark Wigley, 'The Architectural Brain'

Or to put it in historical terms: 'Welcome to Dallas, Mr Kennedy.'

When President John F Kennedy's head exploded across the wide spectrum of communications media operating in 1963, it gave form and momentum to an Electronic Delirium in which a number of conflict-ing effects – textual, environmental and audiovisual – entered into an increasingly unstable relationship with each other. A new architecture came into being: one that connected the central nervous system to a shifting environment of physical and electronic structures, buildings and networks. There would be little to separate media news reports from traffic flows and concrete underpasses, car rental agencies and book depository buildings, mail-order weapons catalogues and tour-ists with cine cameras. 'And suddenly,' Archigram's Warren Chalk announces from the pages of the *Architectural Forum* for October 1966, 'the medium is seen to be more important. Architecture will no longer be concerned with individual buildings or groups of buildings, but with forming a permissive environment that is capable of any configuration according to circumstances.' Benjamin's age of mechanical reproduction, in which the ritual is replaced by the political, is consequently thrown into complete disarray: more than forty years of conspiracy theories have helped to take care of that.

35

84

· 3 ·

Chalk's choice of title for his essay already anticipates the new shaping of social space: 'Hardware of a New World'. The changes brought about by digital technology are so profound and pervasive that it is no longer either useful or appropriate to talk in terms of a digital culture but of a Digital Regime: one whose effects are as much economic, social and legal as they are cultural. That it has been possible to refer in the past to a 'digital revolution' implies a process already geared towards the relative stability of a regime. Just days before the opening ceremony of the Beijing Olympics takes place in August 2008, an Economist leader points out that 'the spread of the internet and mobile telephony' has already done more to transform Chinese society than the staging of such an international spectacle ever could. In the age of digital representation, politics is giving way to ritual. What kind of welcome now exists behind the Great Firewall of China?

79

Beyond the software platforms used to process text, images, music and design are the networks that regulate delivery and maintain copyright control. Access codes transform soft spaces into hard ones: anyone who has ever opened either an email or a social media account will appreciate the importance of names and passwords to online security. Much of the keyword blocking technology used by Chinese authorities to prevent access to offending sites is supplied by western companies. Access to online weather reports is similarly blocked in Dubai, thereby ensuring that 'the official temperature rarely exceeds 99 degrees', just below the legal limit set for outdoor manual labourers,

16

even when daytime temperatures regularly climb 'to well over 110', simply in order that its ambitious and exhausting building programme be allowed   to continue. Meanwhile time stands still inside the marble halls of Dubai's climate-controlled shopping malls. 'Days blur with the same electric light, the same shined floors, the same brands I know from home,' observes a correspondent from the The Independent as he wanders through their air-conditioned interiors. 'You see, my son,' Gurnemanz announces in Act One of Wagner's *Parsifal*, 'here time turns into space.' It is a declaration accompanied as much by the sound of moving scenery as it is by talk of shifting consciousness. Wagner himself is keenly aware of this moment's duality. 'The unrolling of the moving scene, however artistically carried out,' he asserts when commenting upon this moment, 'was emphatically not intended for decorative effect alone; but, under the influence of the accompanying music, we were, as in a state of dreamy rapture, to be led imperceptibly along the trackless ways to the Castle of the Grail; by which means, at the same time, its traditional inaccessibility, for those who are not called, was drawn into the domain of dramatic performance.'

This linking of 'dreamy rapture' with 'inaccessibility' anticipates today's equally illusory relationship between the welcome and our absolute need for security. As networks increasingly serve to collapse space, the 'standing up' of panoramas and facades in the wake of this major technological shift represents an attempt to maintain the appearance of continuity. Sports arenas and shopping malls built to pressing deadlines merely provide the means by which we distance ourselves from speedy regime change. The US Administration bunkered down in Baghdad's infamous Green Zone even spoke of how quickly they had 'stood up' new civil or legislative bodies to replace those dismantled in the wake of the 2003 invasion of Iraq. Stage machinery does not effect change: it is change that brings the stage machinery into being. In such a manner is time transformed into space.

• 4 •

The answer to the Riddle of the Ages has actually been out on the street since the First Step in Space. Who runs may read but few people run fast enough. What are we here for? Does the great metaphysical nut revolve around that? Well, I'll crack it for you right now. What are we here for? We are here to go!
– Brion Gysin, The Process   58

Transcending mere events, absolutes tend not to require a history. There is nothing remotely relative about the Digital Regime: the binary coding of zeros and ones upon which it depends ensures that its values are never shaded. Furthermore as the Digital Regime expands, it no longer requires a history but an actual geography: a sense of distances covered, of connection points that link with – and report back on – the trackless ways of a seemingly familiar yet still largely unknown terrain. The collapsing of time and space also collapses our understanding of what is meant by the word 'transport'; and a new metaphysics of communication is forced into being. Or as Microsoft expressed it   25
in their 1994 global advertising campaign: 'Where Do You Want To Go Today?' A confrontational appeal to individual desires, Microsoft's fabulously seductive question dispenses with notions of work in favour of mixed allusions to travel and play, power and command, velocity and distance. The screen becomes a means of transport, not production. Inaccessibility and dreamy rapture take over the desktop workstation, transforming it into a magic carpet. No longer forced to outstay our welcome in the expectation that we might actually be called upon to do something, we are finally 'here to go' – to use a phrase originating from deep inside the Electronic Delirium of the 1960s. Such extreme mobility must appear unencumbered or it will not exist at all. By 2009 Microsoft is marketing the cross-platform features of its newest operating system under the slogan 'Life Without Walls'.

Defining and regulating accessibility and security within the Digital Regime is the Network: the summation of all the trackless ways across time and space. 'Inasmuch as it is a network,' observes

22   Professor Wigley, 'it is a whole world, a complete spatial system: you cannot simply be inside or outside of it. It is a landscape without an exterior. The operational principle is redundancy. There are always multiple pathways between any two points and multiple options being activated at any one time.' The Electronic Delirium of the 1960s expressed itself in terms of decentralised systems and simultaneous happenings: it presented history as the unfolding of human consciousness in time. The Digital Regime, however, represents its instantaneous

22   unfolding in space over a distributed network of multiple pathways
25   and multiple options. 'Distributed intelligence does not involve an idea moving from one place to another,' Mark Wigley continues. 'The network itself is a brain, a thinking machine, and each thought belongs as a whole, regardless of the particular geography being activated at any moment. Events don't simply happen in space. The space itself is the event.'

· 5 ·

The only other manmade structure to regulate accessibility and security
22   in such a manner is the Labyrinth. Built at Knossos in Crete by Daedalus
49   to house the Minotaur, it was designed to keep the monstrous creature from wandering into the outside world while at the same time allowing in those sacrificial victims selected for it to feed upon. 'Daedalus, an architect famous for his skill,' Ovid recounts in Book VII of Metamorphoses, 'constructed the maze, confusing the usual marks of direction, and leading the eye of the beholder astray by devious paths winding in different directions. Just as the playful waters of the Meander in Phrygia flow this way and that, without any consistency, as the river, turning to meet itself, sees its own advancing waves, flowing now towards its source and now towards the open sea, always changing its direction, so Daedalus constructed countless wandering paths and was himself scarcely able to find his way back to the entrance, so confusing was the maze.'

To explore the mythological dimensions of the Cretan Labyrinth is to discover a place where sacrifice, mutation and technology all meet. The Labyrinth is also a network turned inside out: it is all exterior. Humans find they have no place there for very long. Only monsters can reside in labyrinths; and monsters by tradition are never allowed into human society but remain constantly forced out to its perimeters.

A redistribution of walls in space, the Labyrinth extends itself without end or limit. More importantly, it has no gate and consequently offers no welcome to those who enter. As such it represents a territory without borders that has still to be settled and explored. Redundancy as an operating principle is expressed in negative terms by the Labyrinth's structure: it is simultaneously all link and all node. 'I know of a Greek Labyrinth that is a single straight line,' remarks a character in the short story 'Death and the Compass' by Jorge Luis Borges. 'Along this single line so many philosophers have lost themselves.'

Today the shortest distance between two points is all that separates 'Home' from 'End' on a computer keyboard. If the Labyrinth, hardening and hollowing out the network's multiple pathways, seems gloomy today, it is because we have become blinded by its brightness. In Book 18 of The Iliad Homer describes the Labyrinth as a dancing floor created by Daedalus as an offering to the Minotaur's human sister, the 'lovely-haired' Ariadne, for the golden youth of Knossos to move across in unison. Its intricate layout has been skilfully worked into the design of Achilles's shield by the divine engineer Hephaestus: too bad it cannot protect the Greek hero from his own weakness. The hardest labyrinth to escape is the one you don't realise you're already in: another reason why it offers no welcome to the unwary but only a monster hiding within.

38

74. 53

49

· 6 ·

Where do you want to go today? From the Labyrinth to the dancing floor, this question continues to be posed. Flooding the network, it repeats itself with such speed and frequency that, as with the welcome, we scarcely seem to notice it anymore.

Where do you want to go today? The question represents an inverted form of welcome. What begins as a desire often ends in obligation. Caught inside the Digital Regime, people have to travel. Thanks to the proliferation of media files and portable devices, the world seems to demand it of us. Now a thoroughly networked device, the digital camera no longer presents us with an inverted universe trapped inside a lightproof box but with a bright stream of images flowing across a luminous screen.

Where do you want to go today? The network fosters instability: constantly changing shape and format, the digital image has no fixed area, dimension or destination. It can never resolve itself into a stable form. Each tiny glimpse of life on this Earth gives way ungrudgingly to the next. Protean and vital, connecting image files together, ranking and organising them, sites like Flickr and Instagram, Vimeo and You-Tube, along with their imitators and rivals, constitute tiny planets in their own right.

Where do you want to go today? Our digital trails outline the network's invisible parameters: from welcome to welcome, they are the Labyrinth. A distinction consequently begins to present itself between mere travel and the practice of tourism. The latter is thereby revealed as the systematic application of a specific perspective, no different in its effects from either cubism or socialism. Tourism, in short, has become an abstract philosophy: an art movement. We are all tourists wandering around the Digital Regime, going wherever our cameras take us.

• 7 •

'I hate people going places,' Hollywood producer Jerry Bruckheim-   33
er once remarked during an editing session for one of his television   14
shows. Footage of individuals in transit does not move the story along
in his eyes; the next scene begins only when they have arrived at their
destination. The Prologue to Marc Augé's Non-Places: Introduction to
an Anthropology of Supermodernity contains everything that would be
edited out of a Bruckheimer production. Pierre Dupont, a faceless set
of responses and appetites, is followed by an omniscient narrator as
he takes a taxi to the airport, stops at an ATM, passes through to the   87
departure lounge, shops for some duty-free items, boards a plane and
selects his in-flight entertainment.

What kind of scenario is this? How can it be storyboarded? Monsieur
Dupont appears to go nowhere and do nothing. Published in France
at the same time as Microsoft first asks 'where do you want to go
today?', this simple account of a passenger in transit leaves its protagonist suspended between scenes. Written by a French anthropolo-
gist rather than a Hollywood producer, it is best read not as narrative
but as an objective account of the thoughts running through Pierre
Dupont's own head. Travel isolates the self by keeping it in motion.
Consciousness tends to withdraw from a constantly shifting
environment: confronted with a world of increasingly transitory events,
it takes refuge in becoming reflexive. No wonder Bruckheimer has little
time for people going places. They have internalised his drama. Their
response to modern travel has become cinematic.   15

Welcome to your own personal movie. As Augé defines it, a 'non-place'
is rich in information but lacking all sense of locality, characterised by
such lost spaces as motels, motorways, shopping malls, supermarkets
and airport departure lounges. Non-places are designed to be experienced by a transitory populace of consumers, tourists and corporate
employees: all of those in constant need of being welcomed. Offering
a temporary refuge from the Electronic Delirium, the non-place lies at
the centre of the 'almost unbelievable synthetic landscape spawned

by the jet age' described by George Nelson in his 1967 magazine article 'Architecture for the New Itinerants'. There was, Nelson concludes, 'the possibility of moving all over the world without ever, in a sense, leaving home.' Later that same year *Time* magazine celebrates the emergence of a new executive class of 'Corporate Nomads' whose personal geography is located exclusively within the business world. And so with blithe assurance Pierre Dupont moves through the prologue to Augé's book, the distracted focus of his attention following visual cues while barely noticing them. It takes a certain effort of will to get lost in an airport departure lounge, after all. The entire space is organised to function as one vast movie set. You always know where you are, picking up silent directives on how to sit, stand and comport yourself. One year after 'Architecture for the New Itinerants' appears in print the authors of Archigram 8 neatly reverse the perspective on Nelson's 'almost unbelievable synthetic landscape', identifying in the clearest possible terms the cinematic dimension to modern travel. 'The concept of "place",' they assert, 'exists only in the mind.'

Warren Chalk's 'permissive environment' finds its perfect expression in the continuing development of airport design dating from the 1960s onwards. Neither a building nor a machine, the airport terminal exists instead as a complex set of possibilities that perhaps only Daedalus could fully comprehend. The repeated use of the same components in different permutations represents the transformation of the Labyrinth into a timetabled sequence of passageways and concourses, lounges and desks, stairways and aisles. The airport has become a movie set, 'capable of any configuration according to circumstances', except of course for those which would place it as a permanent fixture in the immediate locality. The introduction in 1976 of an international sign system for the guidance and direction of travellers, outlined by William Pereira in his cheerily titled essay 'A Journey to the Airport', only adds the illusion of continuity to its transitory and unfinished nature. 'To the best of my knowledge,' remarks a pilot in 1979, 'I never landed on a complete airport.' No passenger ever passed through one either.

This is where the next scene takes place, however. Temporary but thickly overlaid with audiovisual prompts, Augé's non-place is the space in which each passenger's self-directed movie unfolds. Self-consciousness is an advanced state of being, which is why people so often find it inhibiting. 'I get self-conscious when I look at a camera,' a young woman remarks in the Starbucks at London Heathrow's Terminal 5. A complex drama of locations and effects, cameras and screens, framed environments and neatly-organised sets, such spaces are best encountered as events rather than locations. Part of a promotional masquerade for Absolut vodka, two blonde women are dressed in diamante carnival masks and floor-length red evening gowns. A man in a plain black mask and a tuxedo glides by in the background. 'So where's the wedding going to be?' one woman asks the other.

Not to focus on anything is the best way to take in this environment. It almost comes as a surprise when someone speaks directly to you. Within the monstrous jointed frame holding up the undulating single-span roof and glass facades of Terminal 5's main departure lounge – designed to be 'dismantled and reconfigured as future needs change', according to its creators – the arrangement of flat screen TVs within the main departure hall offers a series of reassuring perspectives. The hidden geometry of the universe and the hidden geometry of electronics identified by Mark Wigley are simultaneously revealed in the high-speed deployment of HD images and texts, suggesting that both have less to do with conveying a message than with the setting out of coordinates. These rapidly moving advertising displays are going to catch your eye anyway, so you might as well use them to determine your precise location. Like them, you are only passing through, after all.

· 8 ·

It was more than mere coincidence that the architects responsible for some of these fortified terminals had also designed penitentiaries. Both the airport concourse and the cell block used similar kinds of logic. Interior and exterior spaces were under twenty-four-hour surveillance from electronic eyes, motion detectors and video cameras. Both inmates and passengers moved through a similar series of sealed passageways, automatic doors, and narrow checkpoints, where personal screenings were administered with metal detectors and body searches. Only the duration of the incarceration differed.
   – Alastair Gordon, Naked Airport

In an airport restroom a man is talking on his cell phone while standing at one of the urinals: both hands are busy. 'No, I'm afraid in that case I'm going to have to go to the media with this,' he says then pulls away from the urinal still talking and, zipping himself up, heads towards the exit. 'I'm going to have to continue this conversation in a less public place,' he announces and leaves without washing his hands. There is something in 'the frontier meeting of two worlds', Marshall McLuhan suggests in 1967, 'that tends to create the emotion of multitude and bigness'. It is the lack of completeness to be found in the organisation of the average airport that conveys this emotion in cinematic terms. No wonder the man in the restroom is more concerned with issues of privacy than hygiene.

It is impossible to do anything in an incomplete space: you can only ever pass through it. Locality has consequently come to be defined as the area where you can get things done. A traveller standing in line at the check-in is overheard explaining that he had decided to stay in the same area where he bought his new Buick as he figured it would always be 'in the shop' being repaired – which it is. The true frontier is one that separates locality from incomplete space. Crossing frontiers is a way to remind yourself of who you are: only those who remain at home, where they find themselves surrounded by such reminders, have the luxury of forgetting themselves, if only for a moment. You spend more time with humanity while in transit: caught up in 'the emotion of multitude and bigness'.

The massing of people together into crowds tends to take place in a state of crisis, however. The 'man-caused disasters' of terrorism reveal the human masses as the medium through which the joint imperative of panic and security communicates itself. The rapid transport of large numbers of human beings consequently becomes the real problem: not the Orange Alert. 'Asian airports have dusted off heat-sensing equipment they had installed after earlier scares caused by cases of avian flu and SARS to detect sick incoming passengers,' *The Economist* reports as fears of a new global outbreak of influenza start to spread. Meanwhile, *The Sunday Times* reveals, researchers at defence contractor Lockheed Martin are looking into the possibility of placing scanners 'in busy but sensitive public spaces such as airports' to 'alert officials to people whose brain patterns suggested undue anxiety or violent thoughts'. In an incomplete space, information design takes over the entire environment. Guided through invisible barriers, carefully coordinated checkpoints and discreet signage, the endless streams of travellers are processed as so much data. There is no escape from the incomplete space precisely because nobody is looking for one. The system has hermetically sealed itself off from the immediate locality; and the Orange Alert is now accepted as the normal state of affairs. Even so, the experience of watching your luggage being loaded onto the plane from a truck on the runway is not a particularly reassuring one.

• 9 •

The more people are connected with each other, the more the city transforms itself into data: a Bright Labyrinth operating within its own coded economy of flows and drifts. The notion of the city as 'social factory' has gone the same way as the factory itself. 'We should all be persuading ourselves not to build but to prepare for the invisible networks in the air,' Archigram's David Greene announces just as the

22

26

industrialisation of space is about to slow to a halt in the West. An entirely speculative architecture, conditioned by the transitory and incomplete, has emerged to take its place. Real-time information technology, ubiquitous computing, augmented reality, tangible interfaces and interactive spaces are accessed via a shifting series of transparent overlays, reflective surfaces, moving images, fleeting connections and temporary structures. 'We have the opportunity to wire up everyone on the planet,' remarks Internet pioneer and Facebook investor Marc Andreessen. 'Facebook is set to take advantage of that.' The post-industrial city exists as a network of decentralised services activated by coded identities and passwords. The distribution of Wi-Fi nodes and fibre-optic routes is already influencing the choice of potential locations for new real estate in Manhattan, while in South Korea, China and Japan an entirely new generation of 'Ubiquitous Cities' has come into being. As their name implies, these fully wired 'U-Cities' totally surround each individual citizen, completing the transformation of travel into work. Built on a manmade island close to where General Douglas McArthur landed at the start of the Korean War, New Songdo City is set to be one of the most wired cities in the world. According to Jeffrey Huang, Director of Media and Design at the École Polytechnique Fédérale in Lausanne, features will include 'fibre-optics to every home, data-sharing, automatic building and utility management, full video conferencing, wireless access from anywhere, smart-card keys, public bicycles with GPS, and public recycling bins with RFID that will give you credit every time you toss in a bottle'. Despite, or indeed because of, such services, the Ubiquitous City overflows the immediate locality, effectively replacing it. Travel becomes a full-time job, meaning that you can suddenly find yourself anywhere in the world. New Songdo City's developers boast that it will contain 'the wide boulevards of Paris, a 100-acre Central Park reminiscent of New York City, a system of pocket parks similar to those in Savannah, a modern canal system inspired by Venice and convention center architecture redolent of the famed Sydney Opera House'. In other words, the Ubiquitous-City is the incomplete space extended globally. A day will soon come when we may never leave the airport. 'Of course,' Jeffrey Huang adds, 'the question is: will this be the next Brasilia? What about the risk of surveillance and privacy intrusion in a totally wired city? Many of these cities are conceived as large-scale experiments, so are citizens also seen as experimental subjects? Are they giving their consent for that?'

· 10 ·

'I get you. The people have become slaves of probability.'
– Lemmy Caution, Alphaville   78

Access activates services. We behave like commuters without jobs, finding things for our cameras to do. Tourism has consequently become a new form of work. Digital cameras, however, do not take photographs: they generate social media instead. Transformed into a data stream, the digital image has no fixed form anymore: only connections. What was once a unique moment of exposure to light and shadow is today dispersed across a billion or so mobile networked cameras as fleeting points of contact between them. Captured and shared but never completed, this new order of visual information quickly disappears into itself. One online editing tool allows digital images of tourist sites to be emptied of all other tourists, effectively depopulating Times Square, the Valley of the Kings and the Piazza San Marco; another produces 'miniaturised' versions of real-life scenes in which people and objects appear toy-like and reduced in scale.

The production of these fake miniatures entails the selective blurring of an image to simulate a shallow depth of field. As with the bokeh effect – a term derived from the Japanese word for 'blur' or 'haze' – the relationship between figure and ground is radically shifted through   44
this artificial narrowing of the visual plane: those areas of the image no longer in focus now convey the most information. The MIT Media Lab has been exploiting this effect to develop a barcode that can be read at some distance by the camera in a standard smartphone but whose presence remains undetected by the human eye. 'This bokeh-code or Bokode is a barcode design with a simple lenslet over the pattern,'

the Media Lab's Camera Culture webpage explains. 'We show that an off-the-shelf camera can capture Bokode features of 2.5 microns from a distance of over 4 meters.' The density of information contained within the new bokode is thousands of times greater than that of a conventional barcode. 'The goal,' states the Camera Culture webpage, 'is to create an entirely new class of imaging platforms that have an understanding of the world that far exceeds human ability and produce meaningful abstractions that are well within human comprehensibility.' In other words, a service environment that contains codes that can only be read by machines will gradually be replaced by one containing codes that can only be tracked by machines. The growing disparity between these two levels of understanding engenders a new form of instinct: one that transforms the spaces around us, rendering them narrow and indistinct. Essentially a form of programming determined by environmental factors, it is this instinct, rather than 'human comprehensibility', that takes over in the activation of services. Far from being erased from the scene, however, we are encouraged by this process of selective blurring to blend ourselves unobtrusively into the foreground. No surprise then to discover that Jean-Luc Godard's *Alphaville*, a film shot on location in the very heart of the Electronic Delirium, a part of Paris redesigned and repurposed for life in the year 2000, was originally to be called *Tarzan vs. IBM*.

19

· 11 ·

Shopping becomes a leisure activity at precisely the moment you stop noticing the other people around you. Your fellow tourists no longer form part of the scene; common humanity has been left behind at the 'frontier meeting between two worlds'. Malls and supermarkets redefine themselves as social spaces in which we don't encounter people but objects: a new set of frontiers has taken over. You remember who you are – that is to say, *where* you are – the moment your card gets swiped against an item. Quick Response barcodes resembling tiny low-resolution labyrinths can be scanned both vertically and horizontally; an intermediate state between image and code, QR design offers only a transitional series of connections to other data sources. Located outside Düsseldorf, the Extra Future Store shifts the products from its aisles of supermarket shelves to your shopping cart through the agency of RFID chips implanted in each and every item. That is no longer a carton of orange juice before you in the chill cabinet – it's your carton of orange juice. It just told you so.

A network of objects is established. More importantly, a supply chain 85 of perishable information has been set in motion. 'The prices on the 35,000 remote-controlled LCD labels flickering on the shelves rise or fall each night with inventory levels,' *Wired* magazine enthusiastically reports from the aisles of the Extra Future Store. 'Throw that 10-pack of juice boxes into the cart when there are still two pallets in the back room and it could cost you 1.99 euros. But show up after a Saturday afternoon rush and you'd be looking at 2.53 euros.' Meanwhile a team of researchers at MIT's SENSEable City Lab develop the 'Trash Track' project in which electronic tags allow consumers to follow the migra-tion patterns of their household waste through the disposal systems of New York and Seattle, revealing the final journey of such everyday articles as that orange juice carton you recently found so fascinating. The tag helps to calculate the location of each piece of refuse with-in the system and then reports it back to a central server, where the data can be analysed and processed in real time for the public to view 23 online. 'The study of what we could call the "removal chain" is becoming as important as that of the supply chain,' one Trash Tag team member explains. The consumer thus becomes identified as a process-ing point, converting new products into garbage.

Ritual has been replaced in the age of digital representation by process, which is the external expression of instability. Even though the effect may only be illusory, the Quick Response barcode and RFID chip help to transform data into an instantaneous presence: one that can only occur in 'real time'. As people and objects come together to form a continuous self-renewing stream of data, the structuring of space is less concerned with the timetabling of discrete events than the directing

24

of flows. The emergent network of people and objects consequently never resolves itself: the architecture of communication remains indeterminate. Self-renewing and perishable, this visual language is not readable by the human subject but exists only as a service supplied by the scanner or smartphone. All-seeing and subject-less, the electronic eye is not easily led astray by the devious paths of the Labyrinth. It does not look out in any particular direction but merely processes whatever passes before it in order 'to decompose sensed values into perceptually critical elements', in the words of MIT's Camera Culture website.

'A significant enhancement in the next billion cameras to support scene analysis, and mechanisms for superior metadata tagging for effective sharing will bring about a revolution in visual communication,' the site continues. Not prone to dreamy rapture, the electronic eye has helped to establish an interactive environment through which we now wander in a contended daze. It supplies us with detailed maps of the immediate area, updated price guides to the products on display; it even opens and closes the doors for us. The 'revolution in visual communication' celebrated by the MIT Media Lab is predicated upon the object individuating itself through interaction with this subject-less eye, maintaining a semiotic presence so far within 'human comprehensibility' that it marks a virtual return to our first language, which was a purely visual one. Speech is disappearing into an invisible stream of services accessed through an entirely new class of imaging platforms, most of which will be located inside your mobile phone. An ad campaign for Canon scanners shows a group of Masai tribesmen holding up brightly coloured printouts. 'We speak image,' runs the handwritten caption next to them.

Behind the daze lies a last trace of the original delirium; after the revolution the regime busily starts to establish itself. Even though they share the same stretch of walkway, domestic air passengers at Gatwick Airport are kept safely apart from international ones by a simple transfer of barcodes. After disembarking from their flight, incoming travellers have their passports checked and are handed a card with a barcode printed on it. They then proceed along a corridor, which also services a separate stream of outgoing travellers, towards a second checkpoint just outside one of the airport's two main terminals. There they return the card to a security officer, who scans it into his computer; passengers with quick reflexes will have just enough time to see on the checkpoint monitor a digital image of themselves in the act of being handed the card with the barcode not thirty seconds before at the other end of the corridor. 'After checking in, passengers have no choice but to walk through a shopping arcade,' architect Wallace Harrison writes in an early proposal for Idlewild Airport in New York at the very dawn of international air travel. 'Along this route are the well advertised, high revenue-producing stands, shops, food and liquor-vending establishments, and direct access to the newsreel theatre.' A CNN morning news anchor shakes herself awake, as if emerging from a dream. 'Oh I'm sorry,' she remarks to her co-presenter. 'I had a moment of interest in what you were talking about.' The more we 'globalise' our thinking, the smaller the picture gets. Consent is barely an issue. The latest generation of RFID-enabled supermarket checkouts, designed to automatically screen all the items in your shopping cart and instantly swipe your card against the total, closely resembles the scanners currently in operation at airport security stations.

46

• 12 •

'Some man or other must present Wall; and let him have some plaster, or some loam, or some roughcast about him, to signify wall; and let him hold his fingers thus, and through that cranny shall Pyramus and Thisbe whisper.'
  – Bottom the Weaver, A Midsummer Night's Dream

Further connecting the checkpoint with the checkout, a think tank set up by Pentagon director of Net Assessment Andrew Marshall proposes that the US military scrap its rigid command structures and adopt a version of Wal-Mart's supply network instead. 'It may seem odd, to say the least, to nominate a chain store that peddles cornflakes, jeans and motor oil as the model for a leaner, meaner Pentagon,' writes Mike

25

Davis days before the 2003 invasion of Iraq, 'but Marshall's think-tankers were only following in the footsteps of management theorists who had already beatified Wal-Mart as the essence of a "self-synchronized distributed network with real-time transactional awareness." Translated, this means that the stores' cash registers automatically transmit sales data to Wal-Mart's suppliers and that inventory is managed through "horizontal" networks rather than through a traditional head-office hierarchy.' In purely technological terms, the think tank's recommendation proposes that precision weapons be replaced by information design in the war against terror. Structure becomes subordinate to connectivity, and strategy to more flexible forms of interaction. Assemblages of people and machines, images and impulses take shape in real time, continually renewing themselves in the desert areas surrounding Baghdad. None of these individual items can exist in isolation from each other, however; and by linking them in real time the network distracts us from noticing its presence. The question of who – or what – gets connected is nowhere near as important as the point of connection itself.

As the ratio of speed to distance alters, the network collapses all established concepts of space and time. The Labyrinth ultimately disappears into itself, drawing attention away from its resolution, which is the line connecting its entrance to its centre. 'I've become very curious about the maps people have in their minds when they enter the internet,' writes Kevin Kelly in the introduction to his Internet Mapping Project. 'So I've been asking people to draw me a map of the internet as they see it.' Viewed online, the results transform personal geography into narrative. The true network, however, seems to have completely disappeared. 'Not everybody has the idea of Internet as an interconnected network,' writes Professor Mara Vanina Osés of the University of Buenos Aires in her preliminary study of the responses to Kelly's call for maps, 'and even when they see themselves as the center they don't see the relationship between the other items of the net. This make us think if we are truly prepared for the, already here and almost saying goodbye, web 2.0, the web of interactions. People use it but apparently they don't think it/feel it.'

What we don't think or feel has already been worked out in advance for us. Averse to 'people going places', film and television dispense with narrative, replacing it with a series of variations on the same set of social assumptions; dialogue, persona and action are, as a consequence, reduced to mere behavioural responses. Behaviourism is in turn exposed as nothing but the representation of agency: a preconceived pattern of stimuli and responses from which the promptings of the individual imagination have been removed. Just as allegory and symbolism were used by craftsmen in the Middle Ages to direct thought by material means towards the immaterial, so information design has rediscovered narrative as a powerful environmental factor, giving form to that which has so far been neither thought nor felt. Owing his livelihood and status to the skilled crafts of the Middle Ages, Bottom the Weaver in Shakespeare's *A Midsummer Night's Dream* is right to insist that 'some man or other must present Wall'. Although they do not 'think it/feel it', he and his small band of players are themselves caught up inside the complex machinery of a stage play. Their lines have already been written for them. However, in place of the bull's head traditionally associated with the Minotaur, Bottom is forced to wear that of an ass in order that he can play his part to the full.

From Wall to Wal-Mart: by entering the Labyrinth we become a part of it. 'I think they're at the beginning,' a woman in a supermarket calls out to her partner from the checkout, referring to the location of a particular product. At the beginning of what exactly? Architecture determines narrative by physical means. The aisles, shelves and in-store displays establish their own storyline. 'Shadows walk on, across floor and walls, looking for the way back, obviously finding it, for in the end they emerge,' writes the German artist Olaf Arndt, describing a short video made by 'embedded' filmmakers Richard Mosse and Trevor Tweeten inside a maze of fake streets used for training combat troops stationed in Iraq. 'And outside,' he continues, 'there is literally nothing. A flat sandy landscape as far as the eye can see. Desert.' Set up outside Tikrit in what was once Mesopotamia and is now the 'Sunni Triangle', the Forward Operating Base Summerall is designed to rehearse trained responses to 'razzias in inner city combat zones'. Behavioural models

16

such as these are designed to draw attention away from their exteriors: they encourage the idea that nothing exists outside them. 'The labyrinthine structure of the test site,' Arndt reveals, 'was supposed to prepare the soldiers for the difficulties awaiting them when it came to orientating themselves during operations in closely built up, confusing, and above all unfamiliar residential areas.' The Labyrinth is activated by our presence. First approached by US soldiers via its airport, Baghdad now looks much the same as any other city at night: the main difference being the marking on the pilotless drones patrolling overhead.

106

· 13 ·

'They would prefer to "swarm" the enemy terrain,' Mike Davis observes of the new networked warriors, 'with locust-like myriads of miniaturized robot sensors and tiny flying video cams whose information would be fused together in a single panopticon picture shared by ordinary grunts in their fighting vehicles as well as by four-star generals in their Qatar or Florida command posts.' Intelligence is movement. The transfer of information is also a transfer of energy and power. It forms an image. The panoptic city, whose walls were first raised in response to sieges and incursions, has become a swirling cloud of unseen data. Recent advances in the real-time processing of information not only ensure that old hierarchical relationships between text, music and image are being dismantled but that newer hybrid forms are coming into being. This is happening quicker than the eye can capture. Nothing is fixed. The distinction between process and reproduction is rapidly being eroded. The concern is no longer with 'print' but with streams of text, typography, illustration, animation and graphics: printing merely becomes a single moment of resolution that can be infinitely delayed if need be.

20

All design is information design. As platforms shift into each other and design becomes more about the manipulation of sound and vision, language and movement, the standard disciplines and techniques blur together; forms become little more than confines, restrictive and anachronistic. Information therefore has no choice but to exceed its own limits, overwhelming the senses to the point where, no longer able to cope with its presence, they are transformed into information as well. 'Your data is 7 pounds 3 ounces,' reads the copy next to the picture of a newborn baby in an advertising campaign for Hitachi hard disk drives. 'Your data wants to be an astronaut,' runs a further entry in the series, followed by 'Your data runs a six-minute mile', 'Your data loves salt', 'Your data is lactose intolerant', 'Your data is ruler of the underworld' and 'Your data sleeps well at night'. Data is a given, forming structures, yet it also comes to us as a stream. In its lack of distinction between sensory inputs the Digital Regime also marks the rise of a distinctly post-human condition.

19

Projections of how cities operate and what their dimensions might be are becoming increasingly temporary in the Digital Regime. All that will soon remain for you to consider is the open question of security and access, both of which are entirely to do with the transmission and redistribution of information. Your smartphone will do the rest. What may startle you by its absence from this picture, however, is the emotional violence so easily accessed in the past through film and television, the constant interplay of sex and death that tends to bloody the most networked landscapes. That which can be defined has no history because it is an absolute; and absolutes tend to exclude one another. There is no communication between them. Not even the Minotaur is free to leave his Labyrinth. Design still seems to hold itself aloof from the grimier exertions of fine art in the twenty-first century: by limiting itself to purely aesthetic considerations, design inevitably shies away from this problem. It refuses to pull the trigger.

25. 21

In the Digital Regime, design is no longer able to render itself transparent in the translation of modern thought. Giving shape to what only exists in the mind, design is modern thought. In 1986, as the Digital Regime started to establish itself, graphic designer Craig Reynolds created a program to simulate the flight patterns of birds. In observing how a flock of crows responded as a unit to a single command he noticed that no one bird was in command: each crow tended to respond only to those nearest it. He therefore programmed his digital 'Boids' to follow three simple rules: steer to avoid crowding local flock mates; steer towards the average heading of local flock mates; and steer to move towards the average position of local flock mates. 'These rules,' observes Japanese urbanist Akira Suzuki, 'which are more negative than positive, precautions rather than precepts, entail no large view of the flock's purpose, nor do they relate to its conduct – that is to say, they are not analogous to an ideology. They merely prescribe the relations between neighbouring crows.' Reynolds discovered that, by following these three algorithms, the individual crows functioned together on the computer screen as a flock, splitting and regrouping to avoid simulated obstacles. Any bird who missed a turning or hit an obstacle would instantly rejoin the flock after a few seconds' confusion. 'This provides,' Suzuki continues, 'a suggestive metaphor which may help us to understand and describe certain urban phenomena – not the structures of the city, whether concrete or ephemeral, nor its elaborate networks of cultural and commercial activity, but the behaviour of the people and things which flow through it.' Narrative has re-emerged in information design as a navigation device. As consumers of networked sites and services wander through the Labyrinth, responding to positive and negative stimuli, the flock and swarm are exposed as models for intelligence conceived in terms of movement. The network, as Mark Wigley points out, is the brain, space is the event. Maintaining an awareness of this condition is another matter altogether. Shakespeare, so far as Bottom the Weaver is concerned, does not exist – not even the most remote possibility of his existence is considered. Akira Suzuki remains optimistic about that one lost 'boid' who is briefly separated from the rest of the crows. An unexpected moment of recognition has occurred: one that is brief, instantaneous and fragile. For one single moment this solitary 'boid' discovers it is caught up in a relationship with the Labyrinth before forgetting again and promptly being welcomed back into the flock.

The Future is what Happens after You're Dead

'The printing press, the machine, the railway, the telegraph
are premises whose thousand-year conclusions
no one has yet had the courage to draw.'

Friedrich Nietzsche
The Wanderer and His Shadow

· 14 ·

In January 1996, as we approach the edge of a new millennium, Wired publishes an exclusive interview with Marshall McLuhan in which he discusses at great length the effects that the new digital technologies are having upon our most basic perceptions. He emphasises the importance of promoting inefficiency among the business communities of the twenty-first century, argues that 'the product promotes the consumer' and dismisses cyberpunks as mere 'sentimentalists'. In fact, the author of *Understanding Media* seems remarkably alert and well informed for someone who has actually been dead for more than fifteen years. As citizens of the Digital Regime, we have grown accustomed to such intrusions. It was McLuhan himself, after all, who noted that we only catch glimpses of the present in the rear-view mirror as we reverse boldly into tomorrow. So where is this new voice coming from? Does it emanate from the past or the future? Or have our perceptions of progress and its discontents been so compacted over time that the future has finally become the rear-view mirror? And is it still possible to change gears with only a phantom hand on the wheel? 'The real message of media today is *ubiquity*,' declares the newly reanimated McLuhan. 'It is no longer something we do, but something we are part of. It confronts us as if from the outside with all the sensory experience of the history of humanity. It is as if we have amputated not our ears or our eyes, but ourselves, and then established a total prosthesis – an automaton – in our place.'

8, 33

32

At the actual hour of his death on New Year's Eve 1980, the man respon-sible for introducing into mass consciousness such flashcard concepts as 'the global village' and 'the medium is the message' has been all but forgotten by the public, each new book selling progressively fewer copies than the last. There is a time in the 1960s, however, when Professor Marshall McLuhan's face features on the cover of *Newsweek*, his name slapped across car bumper stickers, his reputation as a wild predictor of cultural upheaval the subject of *New Yorker* cartoons. An academic who prefers talking with businessmen to his fellow academics, a soberly attired conservative polymath whose perceptions mesh with the colourful sloganeering of youth culture, what plugs McLuhan straight into the Electronic Delirium is his ability to explain television to itself. In an age when telecommunications are still considered primary conduits to the future, this is no small talent. The fact that his explanations are rarely understood does not seem to matter much at the time. 'Marshall McLuhan, what yuh doin'?' Henry Gibson will inquire with deadpan naivety on primetime TV as the media guru approaches the height of his media fame. When it comes to the future, however, it seems you can't kid a kidder. McLuhan is still publishing books long after his death: the last one being *The Global Village* in 1989. 'Tomorrow is our permanent address,' he once commented, although it has come to feel increasingly like one for handling forwarded items only. People talk of being nostalgic for last week, but sometimes this is the only way to find out what is going on. The Digital Regime has no real historical presence; memory is just another word for storage. An archive made up solely of updates lacks all spirit. Reborn posthumously in 1996, McLuhan's comments on how the future affects the past are more persuasive than ever.

2, 68

7, 46

· 15 ·

The end of everything we call life is close at hand and cannot be evaded. Ours is a closed universe, yet we need something beyond it.
— *H G Wells*, Mind At the End of Its Tether

'I spend a lot of time thinking about the future. Mind = blown.' James Eagan Holmes posts this message online prior to mounting an armed attack on a cinema audience in Aurora Colorado, killing twelve and wounding a further fifty-eight during a midnight screening of the latest Batman movie. The future is what happens after you're dead, which also means that we are condemned to inhabit someone else's future: dreamed of and planned for by others in the name of progress. After observing that we glimpse the present through a rear-view mirror, McLuhan immediately adds that 'we march backwards into the future' as if to suggest that even his metaphors are somehow receding into the past, from highway to infantry column. No one can embrace progress without also accepting that the dreams and assumptions of the dead have somehow shaped their present circumstances. There is at least some degree of certainty to this attitude. Everything has gone before. The future becomes a medium through which the past communicates with the living.

7, 36

Over the previous hundred years socially progressive thinkers have successfully transformed historical certainty into an insufferable enthusiasm for placing themselves at the very end of human development. 'Man has reached the limit of his possibilities,' H G Wells proclaims in 1945, convinced that there would never be another generation after that one. Like many of his literary contemporaries H G Wells finds his work tempered by despair at the inability of the general public to advance beyond its means. The cultural logic of progress dictates that the dreamer must be left behind by the dream. 'Man must go steeply up or down,' Wells concludes, 'and the odds seem to be all in favour of his going down and out. If he goes up, then so great is the adaptation demanded of him that he must cease to be a man.' Seen from this perspective, progress is merely history turned on its head. Insufferable enthusiasm subsequently gives way to gleeful anticipation: there is nothing to look forward to except the supreme gratification of being the last. Confusing progress with the future is simply a means of expressing an unacknowledged desire. 'Renunciation and delay in satisfaction are the prerequisites of progress,' Herbert Marcuse notes at the start of *Eros and Civilization*, his philosophical inquiry into Freud's

36

28

27

28   *Civilization and Its Discontents*. 'The future of the future is the present,'
         Marshall McLuhan asserts in 1967, giving a further twist to the proposi-
74   tion. Even the Unabomber cares about the future.

· 16 ·

*'If it wasn't for progress, where would we be today? On radio, that's
where.'*
         – *Bullwinkle J Moose,* The Rocky and Bullwinkle Show, *NBC TV*

Progress is for those who can't quite take in the future. It comes at us
a second at a time, its approach indicated by the changing design of
a car radiator grill, the cut of a T-shirt, the lettering on a stop sign or an
advertising display. Progress becomes the inevitable by-product of
daily life piling up before us, taking away our dreams and assumptions,
encouraging us to view tomorrow as little more than a slightly different
version of today. We have no choice but to allow ourselves to be moved
forward: things change without any outward sign of effort or volition on
our part. We experience this collectively as a series of cultural folds in
history: progress in the second half of the twentieth century survives the
'closed universe' described by Wells through the simple expedient of
dividing itself into a succession of decades. This has nothing to do with
the passing of time, however, and everything to do with the escalating
3, 31   relationship of production to consumption, style to content, design to
perception. 'Suddenly it's 1960!' Chrysler proudly announces when
marketing the 57 Plymouth as part of its new 'Forward Look' campaign.
With the calm assurance of someone asleep at the wheel, the public
starts to adjust its rear-view mirror.

Marketing cars in 1957 as products of the future, 'three years ahead
of their time', serves to promote the forthcoming decade in terms of
the mechanical advances made since 1945. They are, however, only
standing in for that 'something beyond' Wells had seen tormenting
humanity. Styled to look like jet planes and rocket ships, what this new

generation of automobiles is ultimately selling to the public are the electronics hidden inside them. The linear spread of motorway systems after the Second World War, allowing a clear view of the road behind, is already becoming a thing of the past. The complex circuitry of the Electronic Delirium, offering a dense weave of multiple sightlines on an ever-changing present, swiftly emerges to take its place. 'In the early days of autos,' *Wired* reports, 'the biggest behavioural concern was whether windshield-wiper motion mesmerized drivers. When Motorola's AM car radio debuted in 1930, its critics worried about the distraction of Benny Goodman's clarinet. Back then people *watched* the radio as it played.'

12, 30

• 17 •

Viewed in retrospect, the most exciting development to emerge out of the 1960s is the 1950s. While the former came into sharper focus as the decade of the future, separating itself from the latter, so too did the concept of associating each decade of the twentieth century with significant cultural, social or political change. This perception has increasingly become blurred, however. Only in the twenty-first century could an issue of *The Sunday Times Magazine* boast of interviewing 'Joan Collins on the year that they finally killed off Alexis Carrington, surely the beginning of the end for the eighties, in our special 1989 edition.' Even at the simplest level it is no longer possible to identify progress so casually with a decade's specific look, mood or direction.

52

'Almost every crappy blog post, news story and website published in the last twenty years can be brought up with a quick Google search,' runs a *Daily Telegraph* online complaint at the end of 2009. 'But obtaining contemporaneous informtion from previous decades remains difficult, despite projects to bring newspaper archives, books and records online. The result is an internet that gives undue weight to modern ephemera.' Any clear sense of the 'decade' as a complex and protracted social or cultural event left us during the 1990s: the

85

separation of one decade from another ended with the inevitable logic of one century separating from another, one millennium detaching itself from the next. Unable to maintain a historical presence, the Digital Regime continues obsessively to document a moment of profound crisis: its own creation.

· 18 ·

40

We're all living in a science-fiction world today – and why not?
– Kim Stanley Robinson

The weight and momentum with which progress continues to fold the past into the future as the end of the millennium approaches is increasingly expressed in terms of madness, panic and tension. The last change to the Heaven's Gate website is made on 24 March 1997. A flashing 'Red Alert' notice warns that the passing comet Hale-Bopp will bring 'the time for the arrival of the space craft from the Level Above Human to take us home to "Their World"'. The thirty-nine members of the Heaven's Gate collective in California have by then made their decision to meet the approaching spaceship halfway by ridding themselves of their bodily 'containers' in a calm and carefully planned act of mass suicide. 'Planet about to be recycled,' their website warns. 'Your only chance to survive – Leave with us.'

As it draws closer, the year 2000 comes to be regarded with waning confidence. An inevitably recycled moment, having been anticipated for so long, it will mark either the end of an old future or the beginning of a new history. The narrow increments of progress leave no room at all for the impending apocalypse: one does not plan to celebrate the new millennium so much as to survive it. The Electronic Delirium of the 1960s finally gives way to digital mass mania when experts begin to predict that every computer in world, unable to recalibrate itself to the millennial double zero, will simply shut down at the last stroke of midnight, 31 December 1999. Without the necessary data processing systems to support it, civilisation will sink into chaos: elevators will become stuck in tall buildings, communication networks will crash, emergency services will find themselves paralyzed, and commercial airliners will fall in flames from the sky. 'Maybe they're crazy,' an unidentified female member of the Heaven's Gate community reveals in a videotape message recorded just before they take their lives. 'But I don't have any choice but to go for it, because I've been on this planet for thirty-one years, and there's nothing here for me.' Serenely laid out on their bunk-beds, cots and mattresses, the discarded bodies of the Heaven's Gate 'Away Team' bear witness to the fact that the future is finally starting to catch up with itself. Already rehearsed obsessively in the collective imagination, the events of September 11, 2001 play out a nightmare scenario still waiting to happen.

The small cultural folds of the twentieth century ultimately disappear into a far greater 'Millennial Fold' whose influence can be felt throughout the period stretching from 1997 to 2001. The Heaven's Gate dream of leaving the body behind to relocate on some higher unseen level and government warnings of an undetected 'Millennium Bug' ready to reboot the entire planet are both examples of how the future has become a means of perception. They encourage the public to engage with science-fiction fantasies in which the human appears to have reached its absolute limit. Finality is an inexhaustible concept, however. Progress continually doubles back on itself. Meanwhile science-fiction imagery enters the mainstream imagination as a set of epistemological possibilities: robots and space stations, capsules and pods constitute the Art Nouveau of the twenty-first century, facilitating shifts in consciousness by giving them temporary outward form. In an increasingly fragmented culture these images become a meeting point for collective thinking on social change, technological shifts, political upheavals, plagues and disasters.

35

30

• 19 •

I've always been careful never to predict anything that had not already happened. The future is not what it used to be. It is here.
— Marshall McLuhan, TV interview 1970

A widely accepted representation of the uncanny as populist endeavour, science fiction is the articulation of a machine environment in which, as Professor Friedrich Kittler observes, 'media determine our situation'. The Millennium Bug consequently finds a clear response in popular culture, simultaneously placing the mainstream at the edge and bringing the edge into the mainstream. A cliché, after all, is just a trick that doesn't work anymore: one that has grown a little too comfortable with its context. 'I can feel it,' HAL 9000 murmurs distractedly as his circuitry is disabled in Stanley Kubrick's *2001: A Space Odyssey*. 'My mind is going.' Having decided that further human involvement would be detrimental to the smooth running of a manned mission to Jupiter, the supercomputer has been shutting down the crew's life support systems. The lone surviving astronaut elects to disconnect HAL 9000 from the rest of the spaceship, effectively isolating the machine from its sensory apparatus. Dangerously schizophrenic, HAL 9000 reverts to early and repressed memories in order to shore up a failing identity. Hidden behind a calm uninflected voice, HAL's homicidal breakdown anticipates the mass data panic of the late 1990s, particularly when linked to the movie's title, which cannot help but remind the public that the countdown to a new set of big numbers has already begun.

A future that can be fully described has already happened and, as such, is already compromised. In April 1968 McLuhan has to be constantly nudged awake during an advance screening of *2001: A Space Odyssey*. 'It's nice to have the avant-garde behind you,' he remarks afterwards. Compared to the uncertainty of today, the certainties of the future seem sharply etched and clearly defined – and therefore also more ridiculous. 'For the past 3500 years of the Western world, the effects of media – whether it's speech, writing, printing, photography, radio or television – have been systematically overlooked by social

102

13, 50

observers,' McLuhan tells *Playboy* the following year. The future emerges as a negative critique of the past. It is interesting that McLuhan should have chosen a movie so sharply critical of established concepts of human progress to sleep through. Shifted forward one letter, the acronym HAL becomes IBM, whose advertising slogan 'information overload equals pattern recognition' McLuhan is fond of applying to the Electronic Delirium of the 1960s. 'Electrically,' McLuhan states in 1967, 'it is possible to put every book in the world and every page of every book in the world on one desktop. It's the kind of development that seems to make science fiction seem very silly. Science fiction is very far behind what is happening.'

10, 77

· 20 ·

We tend to forget about the future when confronted with science fiction. Glimpsed through the rear-view mirror, it reflects precisely what is happening now. The titles of Marshall McLuhan's first two books would not have looked out of place on the cover of *Astounding Science Fiction*. *The Mechanical Bride* may be qualified by its subtitle 'The Folklore of Industrial Man', and *The Gutenberg Galaxy* by 'The Making of Typographic Man', but both start out by connecting their readers directly with a universe of robots and rockets. Arguing that we inhabit a technological world made up of tools that extend the body, together with our internal and social worlds, *Understanding Media* seems more unassuming in its choice of title, explicitly referencing a volume on literary criticism, *Understanding Poetry* by Cleanth Brooks and Robert Penn Warren. In private, however, McLuhan refers to his third book as *The Electronic Call-Girl*. Its subtitle: 'The Extensions of Man'.

13, 58

Being published in 1964, *Understanding Media* is already positioned firmly within the machine environment. 'In *Fortune* magazine, August 1964,' McLuhan notes in a lecture delivered that same year. 'Tom Alexander wrote an article entitled "The Wild Birds Find a Corporate Roost" The "Wild Birds" are science-fiction writers retained by big business to invent new environments for new technology.' During the 1960s science fiction is increasingly being used to sell everything from automobile and aerospace industries to home furnishings and Hostess Twinkies. Walt Disney's 'Carousel of Progress' supplies the main attraction at the General Electric Pavilion for the 1964 New York World's Fair, while the General Motors Pavilion invites visitors to its 'Futurama Ride'. Even as such speculative architecture is taking shape, the 'Zoom' issue of *Archigram* utilises science-fiction imagery from comic books and pulp magazines in the 'search for radical valid images of cities'. Also appearing to draw inspiration for its title from *Astounding Science Fiction*, Herbert Marcuse's *One-Dimensional Man:* Studies in the Ideology of Advanced Industrial Society appears in 1964, offering a radical critique of how technology does not extend our inner and social worlds so much as subsume them. 'The liberating force of technology – the instrumentalization of things,' runs an important strand of its argument, '– turns into a fetter of liberation; the instrumentalization of man.'

A photograph taken around the same time in a RAND Corporation office shows a chalkboard covered with notes and equations: appended to one side, amid some other papers, is cover art for the Robert Heinlein novel *Starship Troopers*. A certain malign glee accompanies the discovery: the ideological essence of nuclear deterrence is reduced to waging a technological war against giant bugs. When it comes to the future, expectations change. Relocated to Disney World, the 'Carousel of Progress' requires updating in 1967, 1975, 1981, 1985, and 1994, while 'Futurama' is a modernised take on a ride first shown at the New York World's Fair in 1939. Constantly renewing itself, the future becomes an expression of waste and excess, its expression being seen as something tacky and overlooked. Marshall McLuhan has impressed upon him by Wyndham Lewis the idea that the artist is 'engaged in writing a detailed history of the future because he is aware of the unused potential of the present'. The Electronic Delirium is also an energy flow: a vast expenditure that catches up McLuhan and Marcuse, Norbert Wiener and Gordon Park, R Buckminster Fuller and Norman O Brown, John Cage and Andy Warhol, William Burroughs and Tim Leary. *Understanding Media* and *One-Dimensional Man* soon share shelf space with *The*

9, 32

2, 50

42. 33. 42
33. 63

Politics of Ecstasy, Operating Manual for Spaceship Earth, Love's Body, Nova Express and Heinlein's *Stranger in a Strange Land* – thought to form part of Charles Manson's favoured reading. When it comes to wastage, no war is too small.

'Does not the threat of an atomic catastrophe which could wipe out the human race also serve to protect the very forces which perpetuate this danger?' runs the opening proposition to Marcuse's *One-Dimensional Man*: the new technological environment appears backlit by its own inevitable destruction. McLuhan's 'extensions of man' instantly flips over into Marcuse's 'instrumentalization of man'. Tools 'grow out of our structure' Emerson's essay 'Works and Days' argues   39
in 1870. 'The human body is the magazine of inventions, the patent office, where are the models from which every hint was taken. All the tools and engines of earth are only extensions of its limbs and senses.' In 1950, at the very dawn of the cybernetic age, Norbert Wiener declares that 'the transportation of messages serves to forward an extension of man's senses... from one end of the world to another.' The electromagnetic pulse of the Hydrogen Bomb, powerful enough to cancel out completely the Electronic Delirium, functions as a medium of global communication. Sensibility and moment become one: the assassin's rifle and the tourist's movie camera are both trained upon   6, 47
President Kennedy's exploding skull. It's all over in an instant. 'In the   2, 110
future everybody will be world famous for fifteen minutes,' Warhol predicts in 1968, speaking of tomorrow with the kind of confidence which, during the 1950s, had made private helipads, robot housemaids and vacations on the Moon only a matter of time. Anti-war protestors chanting 'The Whole World Is Watching' in the face of police brutality on the streets of Chicago that summer express the same prediction as social and political reality.

· 21 ·

Left to speak for themselves, the conditions speak loudly enough.
Perhaps the most telling evidence can be obtained simply by looking
at television or listening to the AM radio for one consecutive hour for a
couple of days, not shutting off the commercials, and now and then
switching the station.
        – Herbert Marcuse, One-Dimensional Man

As a display of national mourning, all commercial broadcasting, including popular entertainment shows and television advertising, is suspended in the United States from the announcement of JFK's death to his state funeral. Leaving the conditions to speak for themselves transforms them into an environment, which is to say into a set of services; any new medium is revealed through its influence rather than any apparent content. Marcuse had already glimpsed the shifting relationship between content and service in his reading of Freud during the 1950s. 'The primary content of sexuality,' he states in *Eros and Civilization*, 'is the "function of obtaining pleasure from zones of the body"; this function is only "subsequently brought into the service of that of reproduction." Freud emphasises time and again that without its organisation for such "service" sexuality would preclude all non-sexual and therefore all civilised societal relations – even at the stage of mature heterosexual genitality.' So pervasive is such an environment that you only have to switch stations 'now and then' to become aware of its effect.

The temptation is to see all such services as working too efficiently; they determine people's situation too well. Technology as a consequence tends to function flawlessly in McLuhan's percepts; its influence can easily be spotted because there are no discernible aberrations or deviations in society's response to obscure it. Freed from unexpected consequences and unforeseen applications, conditions need only be left to themselves in order to speak loudly to Marcuse's readers. Obtaining pleasure from the media becomes a suspiciously straightforward affair. A god with artificial limbs, as Freud characterises modern man in *Civilization and Its Discontents*, might amuse itself in far more

13, 29

interesting ways. 'When TV is older it will use the cognitive dance of our brain cortex directly via the cathode x-ray tube,' McLuhan makes a point of predicting. 'This may be utterly disastrous, like nuclear weapons...' When asked whether he is optimistic or pessimistic about the future, McLuhan likes to reply that he is 'apocalyptic': a statement that leaves nothing behind it. In science-fiction fantasies at least, robots do not always function as intended, machines are often abused and the planet Mars is shadowed by unknown dangers.   39

Artificial limbs both extend and disable: they certainly instrumentalise the possibilities of obtaining pleasure. The age of mechanical reproduction marks the end of 'high culture', as both McLuhan and Marcuse will have understood it, and the beginning of a society that operates, as Freud observes in *Civilization and Its Discontents*, along aesthetic rather than ethical principles. Culture becomes a text made up of fragments, details and allusions that has been transferred from the printed page onto the immediate environment in the form of billboards, magazine covers, screens and facades. In a mass society conditioned by technology there is no possibility of evading this: even 'high culture' owes something to what is described by Heidegger – with whom Herbert Marcuse once studied – as 'the essence of technology'. Being a German scholar, Marcuse makes a clear distinction between a 'one-dimensional civilization' and 'a two-dimensional culture'. Culture   29
impedes progress; civilisation trails behind it. Both aspire to conditions of permanence; but that which is eternal must renew itself eternally or be eternally obsolescent. Beyond civilisation and culture lies the regime. The logic of the electronic world is stasis, announces Marshall McLuhan: a man not given to using the word 'progress' if he can possibly avoid it. Tribal and electronic cultures share many of the same bodily pleasures, according to McLuhan, because neither is organised in predominantly literate – which is to say, visual – terms. It is a perspective that leaves little room for the more intricate mechanics of progress. 'Mythology,' Marcuse argues, 'is primitive and immature thought – the process of civilization invalidates myth': which, if not an actual working definition of progress, is certainly symptomatic of one. Taking the long view is always necessary where the manufacture of gods is concerned.

· 22 ·

As educators and responsible citizens, we have to inquire whether we choose to pay the price for a technological change which not only substitutes multiple spaces and times for our long-held Euclidian world, but which also pulls the rug out from under all the legal, political and educational procedures of the past three thousand years of the Western world.
  – Marshall McLuhan, 'Technology, the Media and Culture'

No one chooses to be a prophet: future events tend to interfere with all of your assumptions, thrusting them back upon you. To be without honour in your own country is to have clearly failed to recognise this. 'That's how I prophesy,' pronounces McLuhan in the 1970s. 'I look around at the effects and say, well, the causes will soon be here.' Confidence in the future had already begun its steady decline – until finally a conference of futurologists meeting in New York in 1999 to discuss life in the twenty-first century find they can only agree upon one thing with any degree of assurance: the 'paper-free office' is still some way off. 'Where's my jetpack?' subsequently becomes a common baby-boomer complaint, suggesting that their future, like their past, can only ever have been televised. 'These consequences ensue simply from stepping up information to the speed and level of the electronic,' McLuhan remarks in 1960. 'As is well-known in information theory, as information levels rise, not only one kind of natural resource becomes substitutable for another, but any subject-matter divisions also disappear. We are left confronting primal lines of force and development in a unified field… There is no history, for all time is now.' Merely a repeat in another form, prophecy easily adapts itself to network TV schedules.

The paradoxical logic of progress, as McLuhan understands it, can only be resolved in spatial terms: glimpsed through a rear-view mirror at some remove. Repetition becomes important, as does the expansion of scale. Ed Ruscha's 'Cinemascope' rendering of words, Rosenquist's billboards, Lichtenstein's giant comic-book panels and Oldenberg's sculptural enlargements of everyday objects all strive to dominate a

89

landscape of highways and feeder roads. Responding to the mechani-   4, 40
cal advances of the twentieth century, each of these works threatens to
become a continuity in itself: a fragment of some vast and inaccessible
separate universe. Things seen at a distance are endowed with a greater sense of order: they become a part of history. 'It is no longer possible
simply to communicate a precise body of knowledge founded on a
rigid status quo,' architect Konrad Wachsmann will note as the Electronic Delirium begins to make its presence felt. With no fixed point
from which to view it, the Labyrinth is an endlessly renewed sequence   5, 40
of unevenly-distributed links and nodes.   4, 64

Commuting, especially in crowded transport systems, reveals the unstable condition in which the connected citizen now finds himself. 'The
subway during the evening rush hour,' Marcuse notes. 'What I see of
the people are tired faces and limbs, hatred and anger. I feel someone
might at any moment draw a knife – just so. They read, or rather they are
soaked in their newspaper or magazine or paperback.' We are made
more fully aware of media's effects when in a crowd – even more so   8, 26
as the flow of people becomes increasingly wired into itself. With the
introduction of multi-service handheld devices such as smartphones,
tablets, MP3 players and e-readers this 'soaking' of the senses to the
point of saturation is even more apparent during rush hour. The effects
are already here. The imbecilic audiovisual noise of these handheld services encourages us to remain blissfully unaware of our status as *their*
content. They work us, our choices being limited precisely to the connections they supply. Who or what are the isolated 'faces and limbs' of
this rush-hour crowd finally connected to? Marcuse offers a glimpse of
the violence hidden within this condition. 'And yet,' he comments of
his fellow commuters, 'a couple of hours later, the same people, deo-
dorized, washed, dressed up or down, may be happy and tender, really
smile, and forget (or remember). But most of them will probably have
some awful togetherness or aloneness at home.' You only have to contemplate the number of people listening to music on the way to and
from their jobs to realise that it can no longer be seriously considered
as a force for changing lives.

· 23 ·

Jack Ruby shot Lee Oswald while tightly surrounded by guards who were paralyzed by television cameras. The fascinating and involving power of television scarcely needed the additional proof of its peculiar operation upon human perceptions.
    – Marshall McLuhan, Understanding Media

Behaviourism concerns itself primarily with the rationalisation of external factors: B F Skinner's Operant Conditioning Chamber enforces behaviour patterns through the repetition of negative and positive stimuli. Initially designed to train animals to follow certain procedures by the systematic administration of electric shocks or food pellets, the 'Skinner Box' maintains its authority by locating itself outside the subject. All-embracing yet self-contained and separate, the device exists purely as a series of effects and influences. Information overload equals pattern recognition, which registers as temporary paralysis. Surrounded by lights and cameras in the enclosed space of a city jail basement, Oswald's guards never knew what hit them. 'Ordinary people, simply doing their jobs, and without any particular hostility on their part, can become agents in a terrible destructive process,' writes Stanley Milgram of his experiments in the social psychology of obedience conducted at Yale during the early 1960s. 'Moreover, even when the destructive effects of their work become patently clear, and they are asked to carry out actions incompatible with fundamental standards of morality, relatively few people have the resources needed to resist authority.'

More interesting than its actual findings is the staging of the infamous Milgram Experiment itself. What was originally advertised as a 'study in memory and learning' turns out to be a carefully rehearsed drama for two performers and one naive participant who remains unaware that he or she is the actual subject of the operation. An early form of reality TV, the experiment requires props and plot devices in order to function effectively. The participant does not know that his or her role as 'teacher' has already been fixed beforehand. The electric shocks are faked, and the 'pupil' to whom they are administered is not suffering from a heart

77

104

condition. Similarly the 'experimenter', with his clipboard and grey lab coat, can only choose his responses from a limited number of options. As the pupil makes more and more 'mistakes' in the memory exercise being conducted by the teacher, the voltage of each shock is increased by degrees until it finally reaches fatal levels. The setup itself constitutes a fourth protagonist, stepping up the action. A medium by which the teacher, although unable to see him, communicates directly with the pupil, it allows electric shocks and cries of pain to be exchanged with each other.

Whatever these proceedings tell us about obedience to authority, they are played out in a behaviourist universe, according to a preconceived pattern of stimuli and responses from which the agency of the imagination has been removed. The Milgram Experiment's 'shock box' is simply Skinner's Operant Conditioning Chamber turned inside-out. Behaviourism can only be a representation of agency, however. 'The extent to which this civilization transforms the object world into an extension of man's mind and body makes the very notion of alienation questionable,' Marcuse states. 'People recognize themselves in their commodities; they find their soul in their automobile, hi-fi set, split-level home, kitchen equipment.' With its pushbutton services, the suburban home is revealed as just another version of the Operant Conditioning Chamber. 'And if the individuals are pre-conditioned so that the satisfying goods also include thoughts, feelings, aspirations,' Marcuse concludes, 'why should they wish to think, feel and imagine for themselves?'

Commercially introduced on the iPhone and iPad, the touch screen adds a degree of interactivity to the relationship. Pleasures that once had to be scheduled in advance, necessitating a timetabled repetition of events, are now available 'on demand'. The concept of 'on demand' services, however, is the reverse face of 'real time', locating it firmly   11, 69 within the Digital Regime. Simultaneously transparent and opaque, 'on demand' pleasures work against the principle of delayed satisfaction through which progress operates. 'Everything in one touch,' is the sales slogan for Samsung's 2008 generation of smartphones. 'Projecting current trends,' Marshall McLuhan tells *Playboy* in 1969, 'the love machine would appear a natural development in the near future – not just the current computerized datefinder, but a machine whereby ultimate orgasm is achieved by direct mechanical stimulation of the pleasure circuits of the brain.' At which point paralysis sets in.

• 24 •

It is an outdated and surpassed culture, and only dreams and childlike regressions can recapture it. But this culture is, in some of its decisive elements, also a post-technological one. Its most advanced images and positions seem to survive their absorption into administered comforts and stimuli; they continue to haunt consummation of technical progress. They are the expression of that free and conscious alienation from the established forms of life with which literature and the arts opposed these forms even where they adorned them.
     – Herbert Marcuse, One-Dimensional Man

Back in the 1960s, things move too fast for anyone to pay much attention to what McLuhan is actually saying. His celebrity status remains directly dependent upon the degree to which he is systematically misunderstood, especially by the media. 'I have no theories whatever about anything,' he writes to one detractor. 'I make observations by way of discovering contours, lines of force and pressures. I satirize at all times, and my hyperboles are as nothing compared to the events to which they refer.' This lack of a theoretical system or established set of perceptions works both ways. To understand McLuhan is inevitably to have misinterpreted him: never to have read him is to have studied him well. TV, magazines and advertising will take care of the rest. 'As an investigator,' he reveals to *Playboy*, 'I have no fixed point of view, no commitment to any theory – my own or anyone else's. As a matter of fact, I'm completely ready to junk any statement I've ever made about any subject if events don't bear me out, or if I discover it isn't contributing to an understanding of the problem.'

Look beyond the satire and hyperbole, however, and it quickly becomes apparent that McLuhan joins with Marcuse in striving to identify something that operates through our socially-organised reality solely in order to perpetuate itself. 'A mass medium is one in which the message is directed not at an audience but through an audience,' asserts McLuhan in a comment that meshes neatly with Marcuse's observation that 'domination perpetuates and extends itself not only through technology but as technology, and the latter provides the great legitimisation of the expanding political power, which absorbs all spheres of culture.' Every medium creates an environment of services, allowing human experience to be infiltrated by technology without it being noticed. Culture is the turbulence generated by this process. Common language, Marcuse argues, is quite useless in fully articulating this state of affairs, as it has been handed down to us from TV, radio, cinema, bestsellers and adverts. Our fears and desires, what we love and hate, are consequently described in the same terms as 'our automobiles, foods and furniture our colleagues and competitors'. They represent 'the factual ingression of society into the data of experience'. Or, as McLuhan notes: 'The product matters less as the audience participation increases.' In linking these two statements, we are connecting Herbert Marcuse the media theorist with Marshall McLuhan the radical philosopher.

11, 26

57

In the Digital Regime this level of human participation is approaching the total. Any public scene becomes a drama, thanks to the welcoming intrusion of smart media. 'Rupert, Rupert,' a woman loudly confides to her BlackBerry on a crowded commuter train, 'I love you. I married you. I'm *not* lying to you. I wandered out to get a cigarette.' Among the technologically adept youth, the mobile phone becomes an essential prop for this sort of emotional exchange, its presence indicating the complete involvement of the performers, allowing them to act up for a television camera crew that isn't actually there. The relationship between organised and spontaneous behaviour becomes problematic: 'reality TV' has seen to that. The cause hides itself in its effects. We are offered only the briefest glimpse of what is going on. 'The role of the artist,' McLuhan declares, 'is to create an Anti-environment as a

43   means of perception and adjustment. Without an anti-environment, all environments are invisible.' Marcuse quotes from Brecht writing on the 'estrangement effect' in modern theatre to indicate some of the shifts in perception involved: 'That which is "natural" must assume the features of the extraordinary. Only in this manner can the laws of cause and effect reveal themselves.' How else can design function in the

91   Information Age? 'The "estrangement effect,"' Marcuse continues, 'is not superimposed on literature. It is rather literature's own answer to the threat of total behaviourism – the attempt to rescue the rationality of the negative. In this attempt, the great "conservative" of literature joins forces with the radical activist.' Or Marshall McLuhan suddenly finds himself connected to Herbert Marcuse.

· 25 ·

He didn't think it was funny. He put in his strip just exactly what he saw around him. But our trained incapacity to relate one situation to another enabled his sardonic realism to be mistaken for humor. The more he showed the capacity for people to involve themselves in hideous difficulties, along with their entire inability to turn a hand to help themselves, the more they giggled. 'Satire,' said Swift, 'is a glass in which we see every countenance but our own.'
    – Marshall McLuhan on Al Capp's 'Li'l Abner'

No technology, in and of itself, can provide an anti-environment. Put simply, the network will not save you. 'The more blatantly irrational the society becomes,' Marcuse asserts, 'the greater the rationality of the artistic universe.' Satire, exaggeration and paradox, expressions of 'a knowledge broken', offer an escape from the numbing effects of established discourse. With all historical models of progress, together with all theoretical systems for apprehending them, rendered unworkable, you have to take what you can get. Our enemies, to paraphrase McLuhan, are now working tirelessly on our behalf. 'Why don't you read newspapers?' News Corp chairman Rupert Murdoch is said to have asked Google founders Larry Page and Sergey Brin, mistaking the cause for the effect. 'Understanding culture is our business,' boasts an HSBC promotional slogan in July 2009, running the titles of two McLuhan volumes together and thereby mistaking the effect for the cause. Seen from the perspective of the future, things only make sense when viewed in retrospect. There have always been ghosts in the system.

30

McLuhan's posthumous interview with *Wired* in 1996 is conducted via email, following a series of anonymous postings routed onto a popular US mailing site. Who is answering the questions remains a mystery, however. If the poster is not McLuhan himself, the magazine concludes, it must have been 'a bot programmed with an eerie command of McLuhan's life and inimitable perspective'. In this, the least anonymous of ages, personal identity is a destination: an email address, a function key on a cell phone. A form of free-market reality, defined by the terminals accessible in our immediate environment, has come into being since McLuhan drew his last breath. Authenticity is no longer an issue. When it comes to the future, theories don't allow for much reaction time. Confronted with an unimpeded flow of data travelling at the speed of light down a seemingly endless network of fibre-optic cables, they can only slow you up. McLuhan's 'inimitable perspective' is precisely what's required under such circumstances. Creativity requires confidence and a willingness not to understand, both of which are lost in an instant.

In the meantime, those who do not learn from the future are condemned to repeat it. The 'Singularity', an exponential leap in machine intelligence currently predicted to take place by 2030, is first discussed by statistician I J Good as early as 1965. Similar constructs, such as the state of 'Future Shock' and the coming of a technological 'Third Culture', also date from the same period. At age seventeen, chief exponent of the Singularity Ray Kurzweil appears on the CBS television show *I've Got a Secret* in 1965, playing music composed by an electronic computer that he himself has built and designed. Compared to the future, nothing is more uncertain than the present. In Kubrick's *2001*, as the last of his circuits are disconnected, HAL 9000 sings a song taught to him by Mr Langley, his 'instructor' in 1992. It turns out to be the

4, 41

77

chorus to 'Daisy Bell', first sung by a specially programmed IBM 7094 at Bell Laboratories in 1962. The difference between then and now is what separates notions of progress from those of anticipation.

Just as all fiction takes place in an artificially distended present, creative thought anticipates an endless future. The smart thing, therefore, is to assume that everything you do will one day come under the close scrutiny of intelligent machines. Judgement seems certain: forgiveness perhaps less so. Like a phantom voice coming in over the wire, the future is back – and this time it's personal. Your way of thinking and feeling, the way you even carry your body can do nothing but change. The future will no longer be defined by what it will bring, but how it displaces objects, functions and institutions in time. You will have to find refuge from everything you have been accustomed to believe in; otherwise you run the risk of losing yourself in the familiar – which is to say, the accepted. 'I wish I done beauty now,' a female student mournfully confides to her smartphone. Distracted by the categorical imperative of progress, we have allowed ourselves to become hypnotised by the tools of design; the studio is in turn transformed into a refuge for specialised techniques which are rapidly becoming obsolete. That the spectres of sex and death might also prowl through this sterile space seems, under the circumstances, unthinkable.

13, 29

A piece of twenty-first century camp, science fiction plays ceaselessly with the theme of invasion: bodies, worlds and minds are continually being penetrated by powerful external forces. Rockets, robots and aliens are aggressive expressions of the machine environment. 'Your planet has been invaded,' is the warning recorded onto magnetic tape by William Burroughs in the early 1960s as part of an exploratory reading from his science-fiction novel *Nova Express*. It is a message picked up the following decade by the Symbionese Liberation Army, who end their official communiqués to the outside world with the exhortation: 'Death to the Fascist Insect That Preys upon the Body of the People.' John Carpenter later revives the same scenario in *They Live*, where the population of Los Angeles finds itself having unwitting daily intercourse with predatory aliens. Invasion, in other words, is exposed as a mass medium 'in which the message is directed not at an audience but through an audience'. Like the girl at the end of Carpenter's movie, you have yet to realise who – or what – is having sex with you.

One-Dimensional Man ends with an 'unfortunately fantastic' vision: 'the mere absence of all advertising and of all indoctrinating media of information and entertainment,' Marcuse writes, 'would plunge the individual into a traumatic void where he would have the chance to wonder and to think, to know himself (or rather the negative of himself) and his society.' The 'absence of all advertising' in *They Live* reveals only one clear message, further articulated in the work of designer and illustrator Shepard Fairey: OBEY. Thus spoke Andre the Giant, whose replicated image continues to invade the planet, city block by city block, years after his actual death. From Li'l Abner to Andre the Giant, our negative selves stand exposed as carefully-delineated parody, a cartoon version of everything we have allowed ourselves to become. As McLuhan has stipulated from beyond the grave, ubiquity is an attribute of the digitally 'amputated' selves we have been busily constructing online.

'A few years ago,' states Finnish electronic arts pioneer Erkki Kureniemi   73
in Mika Taanila's *The Future is Not What It Used to Be*, 'a computer was able to simulate a fly's brain. In 2020, a mouse's brain, in 2040 a human brain, and in 2060 the brain of all humans at the same time. Life as we know it will certainly be over.' The digital self is the entire amputation of who we are; our online selves come back to haunt us. In Taanila's 2003 film portrait, Kurenniemi is shown obsessively recording the details of his daily life not just against the ruin of some future posterity but to show the inhabitants of space what it meant to be a human being in the latter part of our earthbound existence. To Kurenniemi, it seems entirely possible that humanity will eventually be little more than a strand of code, a   1, 26
digitised spark of consciousness, neither alive nor dead but desperate to discover for itself the facts of physical existence. 'I register everything with manic precision,' he confides to the camera. 'Video recording, take notes with a cell phone every minute – on the most trivial things: how much a cup of coffee is – what people look like in a particular bar and so on. The history of our times can be constructed.' Meanwhile

4, 51   Microsoft researcher Gordon Bell is using a combination of desktop
           scanner, digicam, heart-rate monitor, voice recorder, GPS logger,
           pedometer, smartphone and e-reader to create a detailed databank
           of his life. Instant history for a last-minute future: its effects can be felt
           everywhere. On 23 March 2003 WKYC-TV News announces that
           'Optimus Prime is heading out to the Middle East with his guard unit
           on Wednesday to provide fire protection for airfields under combat.'
12, 80   A Pentagon spokesman states that they are delighted to have Opti-
           mus Prime – actually a National Guardsman from Ohio who has legal-
           ly changed his name to that of the giant Transformer as a reminder of
3, 65   his missing father – fighting with coalition forces in Iraq. The follow-
           ing month neuroscientist Mitsuo Kawato chooses 7 April 2003, the
69   day when Osamu Tezuka's cartoon creation Astro Boy becomes an
9, 69   active machine, to call upon the Japanese government to commit fifty
           billion yen each year for the next three decades to build a robot with
           the capabilities of a five-year-old child. Like McLuhan himself, the
           future satirises at all times.

Towards a Sexual History of Machines

'I walk among men as among fragments of the future:
that future which I envisage.'

Friedrich Nietzsche
Thus Spoke Zarathustra: II, 20

• 26 •

In 1895 at the Maryiinsky Theatre in the Imperial City of St Petersburg a revised version of Tchaikovsky's ballet *Swan Lake* is presented to the public for the first time. Choreographed by Marius Petipa and Lev Ivanov, one of its highlights is the 'Entrance of the Swans' in Act II, which requires all twenty-four members of the female corps de ballet to file out onto the stage in one long line that continually folds back upon itself. Executing a precisely repeated series of steps, each individual dancer copies the movements of those around her until every gesture has become duplicated, restated and superimposed. The effect is comparable to peering through the slots of a Zoetrope, a piece of Victorian parlour magic capable of creating a limited impression of animated movement. Machinelike and serene, enhanced by the dazzling flicker of such strict repetition, human movement has entered the age of mechanical reproduction. Those responsible for bringing this illusion to life, however, are themselves the victims of an even greater illusion. As women, as workers, as members of society, these dancers enjoy few or no civil liberties whatsoever. They are simply bodies: assemblages of parts trained to exist either in motion or at rest within a specified space. As such, they share roughly the same political status as the apparatus used by the Lumière Brothers that same year to exhibit their set of ten films to excited crowds in Paris. The earliest of these, *La Sortie de l'Usine Lumière à Lyon*, considered one of the first moving pictures of any duration to be screened in public, shows

86
22, 46

9, 55   women workers emerging from the gates of the Lumière Factory at the end of their shift. They form a loosely organised stream in which only the most casual and habitual gestures are repeated at random intervals. A stray dog, some horses in harness, plus a couple of solitary men on bicycles occasionally interrupt the steady flow – make what you will of that. What links each dancer's disciplined configuration to the whirring of the Lumières' projector and the silent dispersal of the predominantly female factory workers onscreen is a precise definition of their shared condition, supplied only twenty years previously by the German mechanical engineer Franz Reuleaux. 'A machine,' he declared, 'is a combination of resistant bodies so arranged that by their means the mechanical forces of nature can be compelled to do work accompanied by certain determinant motions.'

And he should know. President of the Royal Technical Institute in Berlin, Reuleaux was a highly skilled political player when it came to science and technology, helping to establish a patent system while serving on many international juries and committees. His coercive vision of machines as 'resistant bodies' through which compulsion and force were to become meshed with the prevailing political economy 'to do work'

25, 57   helped establish the basic code of mechanical behaviour in the industrial era. Reuleaux, in fact, wrote the book on the subject. *Theoretische Kinematik: Grundzüge einer Theorie des Maschinenwesens*, published in English in 1876 as *Reuleaux's Kinematics of Machinery: Outlines of a Theory of Machines*, constituted the blueprint for an entire regime at a time when the steam engine still supplied the basic model for comprehending the patterns and parameters of social change. As steam, like the water-powered engine before it, tends to generate only circular motion, kinematics concerned itself with the manner in which machines convert one sort of movement into another. To this end, Reuleaux set about systematically analysing and classifying new mechanisms based on how they constrained motion, ultimately establishing in Berlin a taxonomic collection of over 800 mechanical models to illustrate his ideas. More importantly, Reuleaux was also one of the first to use an abstract symbol system to represent machines, devising

24, 48   an operational language derived from a 'kinematic pair', by which the specific coupling of parts is denoted. A complete assembly of mechanisms is a sentence made up of such symbolic couplings, effectively setting a functioning chorus line of resistant bodies in motion.

The basic unit of measurement involved in this process is not hard to imagine: the design specifications for the first fully programmable assembly line commissioned by General Motors in 1956 requires that   53
each robot arm have 'the same space intrusion and reach' as a human. Enter the swans; exit the factory hands. The mechanism has superseded its parts. Thanks to Reuleaux's theoretical principles of kinematics, humanity and machines have confronted each other across a narrowing social divide ever since. As a consequence, the sequential translation of movement glimpsed onstage at the Maryiinsky Theatre in pre-revolutionary Russia can also be discerned in political activist Robert Linhart's first impressions of an auto assembly line in France during the early 1970s. 'It's like a long grey-green gliding movement,' he recalls, 'and after a time it gives off a feeling of somnolence, interrupted by sounds, bumps, flashes of light, all repeated one after the other but with regularity – the formless music of the line, the gliding movement of unclean grey steel bodies, the routine movements: I can feel myself being gradually enveloped and anaesthetized. Time stands still.'

• 27 •

The two primary motions are rotation and sexual movement, whose combination is expressed by the locomotive's wheels and pistons.
These two motions are reciprocally transformed, the one into the other.
Thus one notes that the earth, by turning, makes animals and men have coitus and (because the result is as much the cause as that which provokes it) that animals and men make the earth turn by having coitus.
    – Georges Bataille, 'The Solar Anus'

And when time stands still, it turns into space: 'the unrolling of the moving scene' conjured up by Gurnemanz at the end of Act One in *Parsifal*   3, 51 requires rotational mechanics in order that the forest be replaced by the Great Hall of the Grail Castle. 'It is,' Bataille observes, 'the mechanical combination or transformation of these movements that the alchemists sought as the Philosopher's Stone.' Cause and effect form a kinematic pair, casting a spell over the senses: the illusions of the past become the communications media of the future. Marx and Freud both contemplated the fate of human energy – its channelling, governing and suppression – in an economy powered by waterwheels and steam turbines: that is to say, in terms of one motion converted into another. Societies get the illusions they deserve.

15, 46

Seen as engines of social progress, machines tend to exist in a condition of purity: no aberrations or inefficiencies are allowed to pollute the coupling of cause with effect. Like the most carefully staged illusions, they have no choice but to function perfectly. Beyond the interconnectivity of their parts, machines have little discernible inner life of their own: humans are consequently drawn towards emulating their outer display as a form of eroticism. 'It is generally agreed,' argues Charles R Walker in *Toward the Automatic Factory*, published at a time when robot arms were starting to appear on assembly lines in the United States, 'that interdependent motions performed by a group of persons which follow a rhythmic pattern yield satisfaction – quite apart from what is being accomplished by the motions.' To separate satisfaction from productive labour in this manner is to equate eroticism with work as a socially organised activity. We suddenly find ourselves back at the dance again.

44

The rhythmic patterns established by French court dance in the seventeenth century are an attempt to reflect the external workings of the world as a harmonious interdependent whole. The precursor to classical ballet, court dance is based on prearranged sets of movements that are copied by each of the participants, either together or in sequence, allowing them to see their every pose and gesture precisely repeated by the dancers around them. Costumes and music, scenery and masks are also used to enhance the reassuring impression of being an intimately connected part of a smoothly operating universe; some dances even echo the harmonious motion of the stars and planets above.

In an age when representation is still predominantly allegorical, and mirrors remain little more than handheld appeals to personal vanity, court dance permits individual performers to see reflections of themselves within the carefully choreographed workings of the entire ensemble. They can, in short, examine their presence as part of an integrated outside world.

• 28 •

The early engine is a logical assembly of elements defined by their total and single function. Each element can best accomplish its particular function if it is like a perfectly finished instrument that is completely oriented towards the accomplishment of that function. A permanent exchange of energy between two elements may be seen as an imperfection if this exchange is not part of their theoretical functioning. Also, there exists a primitive form of the technical object, its abstract form, in which each theoretical and material unity is treated as an absolute that has an intrinsic perfection of its own that needs to be constituted as a closed system in order to function. In this case, the integration of the particular unit into the ensemble involves a series of problems to be resolved, problems that are called technical but which, in fact, are problems concerning the compatibility of already given ensembles.

— *Gilbert Simondon,* On the Mode of Existence of Technical Objects

Ideally, such a careful arrangement of bodies would remain meshed with an equally careful arrangement of pleasures. Both, however, are mediated by the uneven relationship that exists between the isolated inner self and the interconnected motions of the external world. 'Shortly after the semi-automatic machines were introduced,' observes Jean-Paul Sartre, 'investigations showed that female skilled workers would allow themselves to lapse while working into a sexual kind of daydream: they would recall the bedroom, the bed, the night and all that concerns only the person within the solitude of the couple alone with itself. But it was the machine in her which was dreaming of caresses.' No matter how smoothly this machine runs, the externalisation of such intimacies leads to the inevitable appearance of inefficiencies and imperfections within the logical assembly of its parts. Eroticism, identified by Bataille as 'one aspect of the inner life of man', is reduced by such endless kinetic coupling to a repeated series of flat gestures.

Left to its dreams, the machine in us produces faults and imperfections. What was once carefully hidden and regulated through the formal presentation of the self to the outside world is now externalised to devastating effect. The empty external self, the amputated gestural identity of the automaton, is the final manifestation of the traumatic separation of the self from the world experienced by the developing ego. 'An infant at the breast does not yet distinguish his ego from the external world as the source of the sensations flowing in upon him,' Freud observes in *Civilization and Its Discontents*. 'Originally the ego includes everything, later it separates off an external world from itself.' The externalisation of this inner life is ultimately expressed by the separating out of the senses; whether one accompanies or results in the other, the effect is catastrophic. This destructive separation feeds technological progress: its objects, Marcuse notes, 'are in most cases actually and violently assailed, deprived of their form and reconstructed only after partial destruction; units are forcibly divided, and component parts forcibly rearranged.' Eroticism becomes another amputation: another negative self deprived of an inner life. Self-possession is only possible when we are able to exist entirely within ourselves.

15, 45

15, 59

· 29 ·

In the camera he has created an instrument that captures evanescent visual impressions, while the gramophone record does the same for equally fleeting auditory impressions: both are essentially materializations of his innate faculty of recall, of his memory.
  – Sigmund Freud, Civilization and Its Discontents

A god with artificial limbs cannot maintain its self-possession; having extended itself too far, it is obliged to treat memory as a disability. Memory in turn needs to realise itself through machines, thus cameras and gramophones have helped accelerate the relationship between human cognition and technology until it is impossible to make a distinction between the two. To watch machines in action is to see memory itself as energy and, as such, to see it eroticised; 'the power of the machine,' Marcuse asserts, 'is only the stored up and projected power of man.' The transactions involved, however, are not possible outside an economic system. Bataille traces a history of eroticism in terms of its relation to work, separating both from the reproductive urges of animal sex. Such a separation, he argues, is conditioned by an awareness   25, 55
of death as loss and absence, to which the responses of eroticism and   21, 48
work are markedly divergent. Where work conserves and regulates human energy, eroticism squanders and exhausts it.

It is this storing up and squandering that allow such writers as Freud and Valéry to see in poetic literature the 'language of the absent': an absence felt all the more keenly after Reuleaux introduces his kinematic grammar. Machines, having no choice, are unselective in their accumulated registering of sex and death. When, in the year following publication of *Reuleaux's Kinematics of Machinery*,   43
Thomas Edison first starts shouting nursery rhymes into the mute,   66
expectant mouthpiece of what he would call his 'phonograph', the printed word is effectively deprived of its predominance as the regulator of human thought. Before even the faintest echo of a tune is registered upon a rotating cylinder, an entire culture has lost its mind.   21, 41
The possibility comes into existence of new communication technologies that are no longer obliged to operate through the medium of symbols but can store the physical traces of the world as sound and light.

What remained of the old signifying 'language of the absent' is still to be found in the symbolic ordering of Reuleaux's kinetic sentence; it is also present in Edison's organisation of his workers at Menlo Park into the first industrial research laboratory geared towards the mass production of ideas. This new materialisation of memory by machine is understood instead as the active inscribing of sound and light upon a passive surface. The unselective noise generated by sex and death becomes an imperfection that interferes with the precise workings of this mechanised inscription process. No wonder Edison and his contemporaries reach back to the languages of Ancient Greece and Rome to name their new inventions.

The etymology of both the 'photographic plate' and the 'phonographic recording' recall the silent internal reading made possible in the classical world through handwriting's power to communicate sights and sounds. Out of the order of one world, the confusion of another emerges. Having become mechanically separated, sight and sound go on to forge complex new relationships with each other. 'Speech has become, as it were, immortal,' *Scientific American* declares after hearing Edison's phonograph in operation for the first time. Memories remain the most plausible of illusions, however. Between EMI's 'recording angel', gazing into space while dreamily inscribing the spiral groove of a phonograph record with a pen, and Nipper the HMV dog drawn towards the gramophone horn by the sound of his dead master's voice, a new formality comes into being. Every recorded song in existence has since become a message received from the dead, requiring what James Joyce refers to at the time as 'a change of manners'. Each expression of romantic love comes from another world: one that can only ever exist outside our own heads.

• 30 •

For I is another. If brass wakes up a bugle, it is not its fault. That is obvious to me : I witness the unfolding of my thought : I watch it, I listen to it : I make a stroke of the bow : the symphony makes movement into the depths, or comes in one leap upon the stage.
If the old fools had not found only the false significance of the Ego, we should not now be having to sweep away these millions of skeletons

*which, for an infinite time ! have been piling up the products of their
one-eyed intellects, proclaiming themselves to be the authors!*
– *Arthur Rimbaud, letter to Paul Demeny, 15 May 1871*

There have always been ghosts in the machine: inefficiencies and   25, 62
imperfections, lapses, stutterings and pauses that call out to us from
somewhere beyond our senses. Their presence can be detected in the
dislocated punctuation and fractured syntax the poet Arthur Rimbaud
brings to his two 'visionary letters', written at the dawn of the 'Founding
Age' of industrial expansion during which the mechanical reproduction
of sound, moving images and text was to flourish. 'The poet makes himself a visionary,' Rimbaud reveals to Paul Demeny, 'by a long, immense
and rational deregulation of all the senses. All forms of love, of suffering,
of madness: he searches himself, he exhausts all poisons in himself, in
order to preserve only their quintessence.' To deregulate the senses, to
respond to what lies beyond them, is to externalise them and ultimately
leave them behind. When, at the age of 63, the great operatic soprano
Adelina Patti hears a recording of her own voice for the first time, she
blows ecstatic kisses into the brass horn of the gramophone. *'Ah! Mon
Dieu!'* she trills. *'Maintenant je comprehends pourquoi je suis Patti!'*

With its contents page for a title, Kittler's *Gramophone, Film, Typewriter* is a disaster registered on paper. The absence of 'and' in its suc-   18, 68
cession of media – which might be read as interrupting a potentially
infinite list with the insertion of a premature conclusion – suggests that
the effects of mechanically separating the senses continue to be felt.
New technologies do not create new spaces: they establish new behav-
iours – or rather formalised versions of behaviour – with which to con-   16, 40
front the existing ones, thereby altering them irrevocably. The studied
delivery of the first recorded voices corresponds to the stiff formality of
the poses already adopted for the early photographic apparatus. The
careful enunciation and rigidity of carriage favoured by polite society
enables these new mechanical media to register their presence more
forcefully upon the Founding Age. At the same time, however, *Gramophone, Film, Typewriter* has the inscription of noise at its core. The earliest known analogue recordings are so unselective in what they pick

up that they even mechanically register their own existence. For Kittler, this completes an information loop so all-embracing that generals and poets, psychiatrists and shop-girls became swiftly caught up in it. Sigmund Freud, in particular, deplores the notion that his sessions with hysterics should be recorded, fearing the play of inhibitions that might ensue from having a gramophone in the room. The careful formality with which they might address this new technology would, he suspects, prevent his patients from making the lapses, slips and mistakes that give utterance to their fleeting innermost selves. Not everyone can make the same joyful self-discovery as Madame Patti. Speech has become immortal because, once inscribed, it cannot initially be erased: silence no longer threatens to close over it. The results of this knowledge, however, are as often a mad drunken babbling as a rehearsed stiffness. The god with artificial limbs now suffers from the disability of memory.

56

· 31 ·

Affordable motion picture cameras and sound recording equipment ushered in an era of mass self-consciousness. The recording phonograph forced people to become performers in a far more profound sense than did the camera, so easily are emotions given away by a quaver in the voice, or a slip of the tongue.
– Booklet note, 'One of One' CD release

'There must be something wrong with this, Sally...' The sound of your own voice recorded phonographically takes you out of your own head: everything that was formerly held in by the formal pose, the polite and carefully enunciated remark, is suddenly set free and becomes noise. A new kind of hysteria is registered, one that is frequently drunk, occasionally uninhibited and rarely alone. One recording, probably made in the late 1940s and remarkable for having survived so far intact, features two female voices identified only as Betsy and Sally or 'Sal Boo' on the record's handwritten label. From the slurring of their words, conversational non-sequiturs, repeated phrases, hesitations and off-colour remarks, it is quite clear that both are intoxicated. The phonograph registering their drunken outpourings has just come back from a repairman in the 'record department', but there is some doubt that it has really been fixed; one voice even expresses a desire to 'take this lovely machine and… cram it up that lovely little man's rear-end'. We also learn that it is a Thursday night and that they are both teachers. Profanities and laughter are mixed with insulting nicknames and references to double beds; a loud Bronx cheer is blown into the recording device, which impassively maintains its presence as a silent participant in Sal Boo's orgy, the climax of which only comes when instructions from the machine's own operating manual are drunkenly read back to it.

'Readjust the tone arm to make a good recording…' This closing of the information loop provides a rare moment of clarity in Betsy and Sally's mechanically rendered delirium, although nothing in this particular recording could be mistaken for rational behaviour. The chance to 'hear yourself as others hear you' feeds back on itself as memory: no wonder our recorded archives are haunted by voices of the dead. Being constant reminders of the past, ghosts render memory unnecessary. The world of sound has consequently remained a mysterious one, full of hidden depths and unexplored dimensions. Unlike sight, our sense of hearing allows us to detect what cannot immediately be seen. One distinguishing feature of Richard Wagner's design for the Festspielhaus at Bayreuth is its recessed 'mystical abyss' of an orchestra pit, effectively concealing the musicians beneath the stage in a darkened auditorium, making the composer's music appear to originate from somewhere beyond this world to blend with the voices of the singers. Completed at the start of the Founding Age, this palace of mechanical illusions, where every production detail is controlled right down to the audience's applause, permits scenes of love and death to be played out in a technological setting. Such sensuality, as Nietzsche observes shortly after the opera house's completion in 1876, is wasted upon a generation of listeners and viewers who 'no longer have their senses about them'. Edison's phonograph will make sure that they would never return.

3, 41
16, 43. 52

'Oh, I can't think of anything...' Sight only tells us what is already here, whereas sound lets us know what is coming. One early application for the phonograph contemplated by Thomas Edison is to capture the last words of the dying as a token of remembrance for family and friends. Edison, however, is also half-deaf and therefore obliged to scream and shout into his own phonograph in order to hear himself as others hear him. The mechanical separation of the senses is often shadowed by human disability. Georges Demeny, brother to the recipient of Arthur Rimbaud's 'visionary letter' of 1871, uses a series of photographs, taken microseconds apart, of his face carefully enunciating a selection of common phrases in order that deaf-mutes might copy the movements of his mouth and thus be able to speak out loud. 'The Photography of Speech', as he calls this new technique, is demonstrated publicly in Paris for the first time in 1891 before an audience of deaf-mutes who obligingly scream 'Vive la France' and 'Je vous aime' at the tops of their voices without ever hearing a word.

• 32 •

Ghosts! Cinerecordings of the vivacious doings of persons long dead. The preservation of memory ceases at the edge of the frame (a 1905 hand happened to stick into the frame... it's preserved, recorded in a spray of emulsion grains). One face passes 'behind' another on the two-dimensional screen.
          – *Ken Jacobs*, Film-Makers Cooperative Catalogue No 5

Ghosts can't help themselves: they are always where they have to be, not where they want to be – flickering in and out of our perceptions as a series of hesitations, stutters and mistakes. Ken Jacobs's description of 'Tom Tom, the Piper's Son', a silent nursery rhyme filmed at the height of the Founding Age, suddenly springs into life as the unknown person's hand comes into shot and out again. Something about that fossilised sliver of light captures our attention like a phantom presence fleetingly glimpsed out of the corner of the eye. All our senses are instantly aroused. As Kittler observes, there is no such thing as a 'silent' movie: its

14, 44

visual rhythms are too emphatic. Made in Leeds in 1888 and considered one of the earliest surviving moving pictures, Louis Le Prince's 'Round-hay Garden Scene' offers a fleeting two-second peek into a world of restless spirits. Its extreme brevity requires that the footage be watched repeatedly, bringing an obsessive, kinetic energy to the movements of its protagonists. So self-absorbed do they seem, caught up in their brief moment of striding, passing and circling, that it is difficult to determine whether they are blissfully unaware of the camera's presence or are studiously ignoring it. But then, who knows where to look when three dimensions are suddenly reduced so unselectively to two? Like ghosts, bodies now disappear as they pass 'behind' each other.

The same year Le Prince films the 'Roundhay Garden Scene', Thomas Edison files with the US Patents Office his idea for a machine that will 'do for the eye what the phonograph does for the ear' by recording and reproducing bodies in motion. Two years later, while about to debut in New York with an exhibition of his moving pictures, Le Prince mysteriously disappears from a train on his way to Paris. Edison's patent for a 'kinetograph' and 'kinetoscope' – camera and viewer respectively – is filed the following year, when Georges Demeny is demonstrating his 'photography of speech' technique to deaf-mutes. In 1892 Edison starts building his 'Black Maria', the first movie studio, and by 1894 is producing his first short films for the public. Georges Demeny subsequently loses interest in moving pictures altogether and returns to the passion of his youth: gymnastics.

20, 39

Enter the swans. It is never easy to know where to look when confronted with the unadorned bluntness of early films. Doing 'for the eye what the phonograph does for the ear', cinematography quickly establishes itself during the twentieth century as the 'music of the future': a term derisively applied to the operas of Richard Wagner by the critics of the nineteenth. Very little, in fact, separates Bayreuth's darkened auditorium and invisible orchestra pit from the Black Maria's lightless and soundless tarpaper walls. Like opera and ballet before them, the silent movie displaces the inner voice of its protagonists. Everything occurs outside the head. Everyone finds themselves plunged into the same psychic space. The phonograph and the kinetograph will eventually strip the dramatic stage of all monologues and soliloquies; such interior voices will never leave the printed page. The stream of consciousness remains the preserve of the modern novel and its silent readership until sound and vision are recombined in the first 'talkies', making the 'voice-over' possible. The more self-conscious the public are rendered by the encroaching presence of recordings in their lives, the less likely they are to acknowledge the existence of an interior voice. The drama of the 'all too human' has to be inferred from its involuntary pauses, hesitations and repetitions: we have been on the alert for the presence of ghosts ever since.

• 33 •

Action is what takes place in front of the camera, with the lights turned on, to throw the rest of reality into darkness... Those for whom not to be seen is non-existence are not alive; and the kind of existence they seek, the immortality they seek, is spectral; to be seen is the ambition of ghosts, and to be remembered the ambition of the dead.

20, 40   *– Norman O Brown,* Love's Body

Ghosts, like machines, have no choice but to haunt the present: their actions are traditionally described as being fragmented narratives trapped in time, moving images of eternity obsessively repeated. There is a play of inhibition about such spirits: they are compelled to do what they do. In fact, no new medium has ever really proved itself until it has been used to communicate with the dead. The very first photograph in recorded history was undoubtedly taken by St Veronica, who wiped the face of Jesus on his painful journey to Golgotha, only to find the Son of God's likeness imprinted on the cloth she had used. Her name is thought to be derived from *vera icon*; meaning 'true image', it represents an amalgamation of Latin and Ancient Greek that might have satisfied even Edison. The first photographer, whose achievement is commemorated as the sixth Station of the Cross, captured not just an image of someone about to die but of spirit made flesh as well. No wonder Christ's last physical agony, his final giving up of the ghost, was marked by a sudden darkness at noon.

Where can such chemical changes take place except in a darkroom? Unlike St Veronica, Edison's Black Maria requires huge amounts of natural light to register an image, necessitating that the studio be transformed, via an overhead skylight, into a large darkened suntrap. The Wizard of Menlo Park's incandescent bulb is still not powerful enough to rewrite the darkness. 'Whether the light is being used for brain surgery or baseball is a matter of indifference,' observes McLuhan of this   14, 51
new artificial daylight. 'It could be argued that these activities are in some way the "content" of the electric light, since they could not exist without the electric light.' The ensuing play of light and shadow helps put ghosts to flight, trapping spirit inside the flesh again. Bill Dellenback, the cameraman responsible for filming subjects masturbating for the Kinsey Institute archives during the late 1950s, 'became adept at setting up a powerful cone of light such that the performer/s felt invisible to observation', making for a more relaxed clinical performance. Edison's first film, made in 1894 of his assistant Fred Ott sneezing, had found its convulsive echo.

Like the therapeutic power of modern analysis, as characterised by Herbert Marcuse, electric light transforms the movie camera into a 'cure from illusions, deceptions, obscurities, insoluble riddles, unanswerable questions, from ghosts and spectres'. The inquisitorial glare of Andy   20, 103
Warhol's early 'Screen Tests' is designed to highlight silently every behavioural tic and facial blemish, combining the invasive inquiry of the Hollywood casting couch with that of the psychoanalyst. Both repro-   7, 45
duce forms of media attention. The counterpart to Warhol's brutal gaze can be found in newsreel footage of Rudolf Nureyev's impassive insolence at his 1967 press conference in San Francisco following his arrest during a drugs raid. Confronted by lights, cameras and questions, he calmly shifts his attention from one reporter to the next, refusing to say a word. By then Warhol had filmed people kissing, fucking and arguing, his silent fixed camera reducing such activities to high-contrast images

outlined by the darkest shadows. As with ghosts, we only see what we permit ourselves to see. 'The finished product looks like an industrial film,' runs a journalist's account of veteran sex researcher Bill Masters screening footage shot at the Masters and Johnson Institute during the late 1950s of women masturbating, 'which makes perfect sense: it was part of landmark research to create plodding user's manuals for the erogenous zones.' Until Masters and Johnson introduce a camera lens into the vagina along with a speculum, the precise anatomical nature of the female orgasm remains a mystery. The woman's hand, according to the journalist, is visible within the frame as she stimulates herself above the speculum. 'We forgot to have the girl take her fingernail polish off,' Masters complains.

• 34 •

My next series will be pornographic pictures. They will all look blank; when you turn on the black lights, then you see them – big breasts and… If a cop came in, you could just flick out the lights or turn on the regular lights – how could you say that was pornography? But I'm still just practising with them yet.
    – Andy Warhol, Art News *interview, 1963*

A stereoscopic daguerreotype taken between 1850 and 1852 by Auguste Belloc appears at first glance to be a classically-inspired study of two female nudes in poses worthy of the Academie des Beaux-Arts. The tableau might represent a scene from Greek mythology if one of the classical nudes were not administering an enema to the other with the aid of a brass syringe and a length of rubber hose. One figure looks modestly away while the other gazes at her own reflection in a mirror positioned above the parlour table upon which she is reclining. Neither of them seems to notice the sleek lines of the metal pump positioned by the photographer in the foreground. The double-frame stereoscopic viewer misses nothing, however – not even the household broom leaning against the back wall next to some framed daguerreotypes.

Was light alone destined to touch the photographic plate? Oliver Wendell Holmes registers 'a surprise as no painting ever produced' when
first viewing a stereoscopic image in 1859. He sees it furnishing the
library of the future. That which has been inscribed upon can no longer
be pure, however. 'True civilisation,' writes Belloc's contemporary, the
poet Charles Baudelaire, 'does not lie in gas, or in steam, or in turn-   57
tables. It lies in the reduction of the traces of sin.' Edison's invention
of the incandescent bulb in 1879 sets new ghosts loose in Belloc's
parlour; from this point on the traces of sin are not so much reduced as
obscured by the limits placed upon personal privacy. 'Time took on
the character of an enclosed space,' Lewis Mumford observes of the   55
impact artificial light has upon the modern household, 'it could be
divided, it could be filled up: it could even be expanded by invention of
labor-saving instruments.'

· 35 ·

For the private person, living space becomes, for the first time, antithet-
ical to the workplace. The former is constituted by the interior; the office
is its compliment. The private individual, who squares his accounts with
reality in his office, demands of the interior that it foster his illusions.
   – Walter Benjamin, 'Paris, Capital of the Nineteenth Century'

Suddenly everyone has something to hide. The impact of glass and steel
upon architectural design gradually straightens out the Art Nouveau
line, leading to a reframing of public and private space. 'The shattering
of the interior takes place around the turn of the century in Jugendstil,'   18, 41
Benjamin observes. 'Through its ideology, the latter seems, it is true,   2, 42
to bring with it the consummation of the interior.' To reveal what has
been hidden is also to transform it: transparency quickly becomes an
issue. Office windows arrange themselves into uniform film strips
whose sole content is the human activity taking place within each
frame. Electric lights, typewriters, adding machines and dictaphones
establish new settings and rhythms for the psychological dramas that

are starting to unfold inside each building. From the office to the interior, the results are best understood, however, as a form of mechanised dance.

Reaching the peak of its popularity just as Edison's kinetoscope is introduced to the general public, the harmonograph offers an idealised visual representation of the smooth, repetitive motions associated with machines at the end of the century. The mechanism is simple enough: two or more pendulums moving independently of each other determine the looping arcs traced by a pen moving across a flat sheet of paper. While the harmonograph's fluid, rotary action links the device with the lubricated motion of valves and pistons, the images it produces conjure up the seductive double curves and empty undulating forms that shape the Jugendstil line. Hailed in 1893 by Stéphane Mallarmé as 'an artistic intoxication and an industrial achievement', the *Ballet Lumière* created by American dancer Loïe Fuller translates this same line into butterflies and serpents through the rhythmic combination of light, movement and fabric. Light, not time, becomes the moving image of eternity. Variations on her 'Butterfly' and 'Serpentine' dances are copied for Edison's kinetoscope and released as some of the first 35mm film strips, helping to establish an international standard for still and motion-picture photography.

51

Upon arriving in Paris, Fuller initially adopts the stage name L'ouie, after the old French word for 'hearing'. Like the harmonograph, her dances create a popular form of visual music just as the kinetoscope is silently imposing its own. Fuller's rhythmic loops and swirls are so enhanced by the play of artificial light upon her body that it is closely associated in the popular imagination with Edison's incandescent bulb. The result is a series of symmetrical swirls and vaginal openings, transforming her entire body into a constantly moving sexual vortex. Fuller is worshipped by poets and public alike as an Art Nouveau 'Goddess of Light'. The *New York Herald* writes of her artificially illuminated costume that 'embroidered serpents seem to glide over its surface with ever-increasing velocity until the lights are suddenly turned out and the whirling form of the dancer is lost to view.' What this new play of light is slow to reveal, however, is that bodies arranged so precisely in space no longer have inner lives of their own – except as extensions of machines.

• 36 •

The dynamomoter's slot glistened vertically.
'It's a female,' said Marceuil gravely, '...but a very strong one.'
    – Alfred Jarry, The Supermale   95

In return, machines extend the body to sexual extremes. Mechanical devices give pleasure by first isolating and then intensifying sensory impressions, hardening and hollowing out the flesh to the point of numbness. 'Labor-saving instruments' become libidinal generators, allowing work and eroticism to be redefined in terms of energy and time. How else could a sexual history of machines end? Straps and paddles, pumps and costumes, cameras and vibrators form themselves and their operators into an assembly line of pleasures. The final outcome of this arrangement is a recurring cinematic fantasy.   15, 41
'Mr C A Rotwang requests the pleasure of your company at dinner and   98
to see a new erotic dancer,' reads the printed invitation on the screen.
Purporting to depict life in the year 2026, Fritz Lang's *Metropolis* haunts   43
the popular imagination more as a set of visual rhythms than a coherent narrative. A clear expression of what eroticism has become by the end of the Founding Age, it is anything but a silent movie. Rotwang's 'erotic dancer' is actually a machine, a robot that has cost him his right hand to create; the black prosthetic glove he wears in its place requires – and receives – no further explanation.

A burnished metal apparition in which Jugendstil and modernist lines meet, Rotwang's robot exists at one ambivalent remove from both body and machine. Pleasure and amputation inform its entire being. Just as Reuleaux's kinematic couplings and Loïe Fuller's *Ballet Lumière* have helped translate court dance into silent cinema, so *Metropolis* rewrites *Swan Lake* simply by reassembling its parts. In one, Rothbart, an evil magician capable of transforming women into swans, creates the evil twin of a beautiful woman to seduce a handsome prince. In the other, Rotwang, an evil magician capable of transforming machines into women, creates the evil twin of another beautiful woman to seduce another handsome prince: Freder, the son of the city's ruler. As the 'Father of Industrial Robots' Joseph F Engelberger will later observe: 'A robot becomes more robotic the more it emulates the human being.' A distinguishing attribute of Rotwang's robot is its supreme indifference to any such relationship, however. Sheathed in synthetic skin and smiling crookedly, it can only arouse and incite. In other words: it knows how to dance. To do so, however, it must assume the identity of a woman who, despite being in a silent movie, only knows how to speak.

Maria preaches brotherly love to the downtrodden workers of Metropolis. After Freder falls in love with her, Rotwang kidnaps Maria and sends out his robot in her place. Good Maria is now Evil Maria. The preacher has become the dancer. A product of the silent cinema, Freder's inability to tell the difference between a woman and a machine is easily exploited. 'She is with your father,' Rotwang informs the excitable youth, sending Freder chasing after Maria's robot double. In a highly dismissive review of Lang's *Metropolis*, H G Wells complains that there is 'no scope at all for looking well or acting like a rational creature amid these mindless, imitative absurdities'. No wonder Serge Diaghilev disapproves of the Ballets Russes being filmed. A publicity photograph taken on the set of *Metropolis* in 1926 shows Fritz Lang sitting in front of a modern drum kit, while Brigitte Helm, who portrays both Maria and her evil robot double, poses with a saxophone. The picture hints at what kind of music is playing when Maria's mechanical doppelganger goes into her wild erotic dance, creating chaos all around her while giving nothing away. Its disjointed rhythms may be silent for the moment, yet they utterly confound the logical classification of movement set down by Franz Reuleaux in Berlin fifty years previously. No human body has ever danced like this before.

15, 39

• 37 •

When he opened the valve of his wit, he seemed to follow after the stream of his words without any control of them. It was no longer a person speaking but a machine driven by some demon. His jerky voice, metallic and nasal, his abstract puppet-like gestures, his fixed expression; his torrential and incoherent flow of language, his grotesque or brilliant images, this synchronization which today we should compare to the movies or the phonograph – all this astonished me, amused me, irritated me and ended by upsetting me.
   – C G Gens-d'Armes on Alfred Jarry

Exit the swans; enter Evil Maria. The dance of the human machine can only take place on film – or rather as film. Whether on stage or screen, the self-enjoyment of the dancer exists as series of smiles, grimaces and gestures. The body arranged so precisely in space has no inner life because everything happens on its surface alone. Exit the factory women too. Lang edits Evil Maria's dance into a series of fragmented asynchronous loops so that there is no smooth flowing line to any of her movements. None of them is connected or completed: they are merely repeated. Time and space are manipulated, as the events and images surrounding the dance are juxtaposed and superimposed – the effect is one of delirium bordering on madness, but whose?

54

Freder's shocked response upon discovering 'Maria' and his father in each other's arms is depicted as visual noise: blurred spots of light and jagged lines scratched directly into the film represent the flood of sensations conveyed by mechanically separated organs of perception. 'After an energetic attempt to focus on the sun,' Nietzsche writes in *The Birth of Tragedy*, composed at the start of the Founding Age, 'we have, by way of remedy almost, dark spots before our eyes when we turn away. Conversely the luminous images of Sophoclean heroes – those Apollonian masks – are the necessary productions of a deep look into the horror of nature; luminous spots, as it were, designed to cure an eye hurt by the ghastly night.' Freder's eye, as with all those watching Evil Maria dance, is anything but cured, however. Instead it registers madness and collapse: apocalyptic readings from the *Book of Revelations* are intercut with fragmented glimpses of Rotwang's robot performing her erotic dance in a nightclub before an audience of men equally incapable of telling the difference between a woman and a machine. Their lust for what they see transforms them into a crowded arrangement of single staring eyes in a darkened flattened space. They no longer have any use for the rest of their bodies as a source of pleasure: the robot's power over them is complete. 'For her,' reads a title at the climax to the machine's dance, 'all seven deadly sins.'

Rotwang's 'Evil Maria' and Edison's Black Maria were always destined to meet. The robot's erotic dance, presented as a series of jumps, loops and cuts, meshes precisely with Freder's oedipal delirium. 'For I,' the silent workings of early cinema dictate, 'is another.' In the meantime Nietzsche has already restated this relationship as a simple direct question in an 'untimely meditation' dating from 1874: 'Are there still human beings, one then asks oneself, or perhaps only thinking-, talking-, and writing-machines?'

· 38 ·

This continuous modification of man by his own technology stimulates him to find continuous means of modifying it; man thus becomes the sex organs of the machine world just as the bee is of the plant world, permitting it to reproduce and constantly evolve to higher forms. The machine world reciprocates man's devotion by rewarding him with goods and services and bounty. Man's relationship with his machinery is thus inherently symbiotic... Now man is beginning to wear his brain outside his skull and his nerves outside his skin; new technology breeds new man. A recent cartoon portrayed a little boy telling his nonplused mother: 'I'm going to be a computer when I grow up.'
– Marshall McLuhan, Playboy interview

Rewind the film to the moment when Freder surprises his father and

Maria in each other's arms: they are standing in the father's office surrounded by adding machines, tabulators and keyboards. It is from here   5, 63
that the whole of Metropolis is controlled and a new consciousness
forged. Alfred Kinsey's statistical survey of human sexual behaviour is
only made possible through the early adoption of such computational
devices, processing the data supplied by thousands of volunteers answering intimate questions at the prompting of an identical form. Communication is anything but smooth, however. Kinsey will not trust either
the telephone or the radio and insists upon writing all of his letters by
hand. There is no room on his standardised form for the ill-fitting parts
that lurk behind speech: the pauses, slips of the tongue and accidental
ellipses that reveal the unconscious urges towards pleasure.

Denounced by Wagner to his own physician as a compulsive mas-
turbator, Nietzsche becomes the first 'mechanised philosopher',
composing *On the Genealogy of Morals* hunched myopically over his
Malling Hansen Writing Ball. It takes the introduction of the QWERTY
arrangement of keys to impede human thought sufficiently to prevent
the typewriter mechanism from jamming. Later the novelist Hubert
Selby will use this same keyboard to record the uncontrolled tics,
habitual reflexes and blind instances of rage that constitute the mute
autism of the human soul without the conventional regulation of
punctuation or spelling. Researchers at the Mayo Clinic have
subsequently extended the interface between thought and inscription by placing electrode grids inside the skulls of epileptic patients,
allowing them to type words using only their brainwaves. How else
are thinking-, writing- and speaking-machines expected to amuse
themselves except by generating their own inner noise? A masturbating body is the first ever machine: from early childhood Kinsey practises
a form of urethral insertion to increase production of sensations. The
assembly line now automates itself.

As humanity becomes increasingly integrated with its technology,
memory will replace all bodily senses. A digital camera is not an eye,
however, and should not be regarded as one – an MP3 recorder is
not an ear nor should it be misunderstood as one. They are isolated

functions of memory. Such a distinction means nothing to the industrial robot blindly picking and placing objects on an unseen, unheard assembly line; KUKA, once its principal German manufacturer, has already programmed a robotic arm to function as a transcribing monk, tirelessly copying out the Lutheran Bible onto an endless roll of paper. The inscription is quite flawless. Founded in Augsburg in 1898, as the forces of the Founding Age are fully unloosed upon the twentieth century, Keller und Knappich Augsburg is known today simply by its initials. From mnemonic acronym to blind transcribing arm: the transition offers a positive response to Rotwang's silent cry of anguished triumph recorded on one of the title cards in Lang's *Metropolis*: 'Isn't it worth the loss of a hand to have created the workers of the future – the machine men?'

The Dream of Venus

'The dream takes us back again to the distant states of human culture and gives us a means by which to understand them better.'

Friedrich Nietzsche
Human, All Too Human: I, 30

· 39 ·

The force of the explosion may be imagined when it is recollected that they had to give the car a velocity of more than seven miles per second in order to overcome the attraction of the earth and the resistance of the atmosphere.

The shock destroyed all of New York that had not already fallen a prey, and all the buildings yet standing in the surrounding towns and cities fell in one far-circling ruin.

The Palisades tumbled in vast sheets, starting a tidal wave in the Hudson that drowned the opposite shore.

   – Garrett Putman Serviss, Edison's Conquest of Mars

New eras begin where old ones end. That, at least, is the way it should go. As early as 1844, Ralph Waldo Emerson surveys Manhattan's busy skyline and declares that 'America is the country of the future.' The thrusting skyscrapers of New York and other major cities along the Eastern Seaboard soon become the heroic expression of this vision. Collectively they constitute a citadel of progress; seen in terms of Friedrich Schelling's observation that 'architecture is a frozen music' they celebrate materialism as the highest human aspiration. The towers of New York also share a drowned counterpart hidden deep below the Atlantic waves. The foundations of Atlantis lie in darkness and ruin within the ocean's vast shifting shadows. In 1866, as recounted by Jules Verne, the USS *Abraham Lincoln* sets off for its fateful encounter with

20, 56

32, 58

Captain Nemo's *Nautilus* from the Brooklyn Navy Yard in New York. First published in 1870, Verne's *Twenty Thousand Leagues under the Sea* describes an undersea world of wonders, including Red Sea coral reefs, Arctic ice shelves and, most strikingly, the 'long lines of sunken walls and broad deserted streets' of Atlantis – a 'perfect Pompeii escaped beneath the waters', according to the novel's narrator. Plato's account of an advanced civilisation of engineers and architects perishing in 'a day and a night' thousands of years before recorded history haunts the imagination of the industrial age.

Atlantis has always been an unfinished story with a legendary ending, preserved in the deep waters of unconscious memory. In an age of expansion, trade and empire, its submerged contours remove themselves from the physical confines of established cartography. American congressman Ignatius Donnelly publishes his best-selling account *Atlantis, the Antedeluvian World* in 1882; the following year Krakatoa erupts, throwing out tidal waves and columns of volcanic ash. By century's end there are already sixteen novels in print that deal with the myth of Atlantis. Although her first major work on religion and science, *Isis Unveiled,* contains only one page on the lost civilisation, Madame Blavatsky's *The Secret Doctrine* is a commentary upon 'The Book of Dzyan', a sacred text composed in Atlantis which reveals that humanity is the fifth 'root race' following the Atlanteans in spiritual ascendancy towards more ethereal states of being.

15, 36
21, 72

In 1896 W Scott-Elliot traces a similar line of evolution from matter to spirit in *The Story of Atlantis.* Two years later, the Martians arrive out of a darkening sky. H G Wells's account of London reduced to an uninhabitable wasteland of stone, glass and steel by invaders from Mars proves too much for *The Boston Evening Post,* which publishes a pirated edition of the novel. Entitled *Fighters from Mars: The War of the Worlds In and Near Boston,* it details the summary defeat of advanced Martian technology by Yankee engineering and knowhow. As if to emphasise the point, a sequel is commissioned from astronomer Garrett P Serviss: *Edison's Conquest of Mars,* in which the great American inventor equips a punitive expedition to Mars with 'electric ships' and disintegration rays. A science reporter for various US newspapers, Serviss packs his thrilling narrative with thinly disguised references to Atlantis. The lifeless remains of Luna, an ancient and advanced civilisation, are discovered on the Moon; while the Martians, it transpires, have been busy breeding an Aryan super-race with humans abducted from the 'Vale of Cashmere', a mythical paradise celebrated by the American songwriter Thomas Moore for 'its temples, its grottoes, its fountains as clear, as the love-lighted eyes that hang o'er the wave'. A native New Yorker, Serviss will no doubt have recalled that the Vale of Cashmere also lends its name to an area of Prospect Park in Brooklyn.

· 40 ·

A reinterpretation of human history is not an appendage to psychoanalysis but an integral part of it.
    – *Norman O Brown*, Life against Death

A young boy walks through the underwater tunnel at the Brighton Aquarium, past sharks and brightly lit tropical reefs. These arched subterranean vaults have been open to the public since 1872. 'What a waste of water,' he remarks to no one in particular. Since the time of Abbé Suger of St Denis, the play of light through glass has directed human thought by material means towards that which is immaterial. Allegory and representation have been out of alignment ever since.

The history of astronomy parallels the splitting and specialisation of literary forms. Poetry and history supplied the basis for explaining the universe, the behaviour of matter and the origin of the gods. Celestial models imposed the earliest order upon our behaviour; the precise repetition of events which allows humans to make best use of time and space was learned from the stars and the seasons. 'By value of the rationalization of the modes of labour,' Adorno and Horkheimer observe in their *Dialectic of Enlightenment*, 'the elimination of qualities is transferred from the universe of science to that of daily experience.'

30, 53

63

18, 88   The novel consequently establishes itself as a means of expressing social classification. Science fiction, especially when concerned with life on other planets, is a vestigial reminder of this separation, which may explain its unacceptable status as a literary form.

22, 70   The play of light through glass and water suggests the first stirrings of some new form, squeezed into existence by the pressures exerted upon it. Immersion will always be apocalyptic in nature. An oceanic sense of self continues to haunt us even when we are on dry land. An upheaval has taken place in the landscape; but all that remains of it are the mountain ridges and rock formations. You can read the colossal lines of force running through them. We have become split from our environment. 'The woman penetrated is a labyrinth,' according to
22, 61
33, 57   Norman O Brown. 'You emerge into another world inside the woman.' At what point in their development did our cities start their expansion deep into the earth? Brown also notes the derivation of the word 'fornication' from 'fornix, an underground arched vault'.

· 41 ·

Industrial exhibitions as secret blueprint for museums. Art: industrial products projected into the past.
    – Walter Benjamin, The Arcades Project

29, 53   Endings often leave such pretty debris behind them. From Atlantis to Mars: high culture can no longer be regarded as pre-technological in origin. Like them, it is too closely associated with ruins for that. The submerged canals, temples and harbours of Atlantis, not to mention its palaces with their indoor 'hot and cold' fountains, find their reflection in the nineteenth-century citadel of progress. No city can be considered modern without a showing of glass, wrought iron and industrial heated water. Advanced plumbing, glazing and engineering techniques create exotic public environments such as the zoological garden, the tropical herbarium and the marine aquarium. In Paris the poet Gérard de Nerval strolls through the Royal Palais Gardens leading a lobster on a blue silk ribbon because, as he explains, they 'know the secrets of the sea' and don't bark. Others prefer the company of turtles on their slow drift through the capital. Stranger creatures than these will be found wandering the streets before too long. 'Far away, through a gap in the trees,' recalls Wells's narrator in *The War of the Worlds,* 'I saw a second Martian, as motionless as the first, standing in the park towards the Zoological Gardens, and silent. A little beyond the ruins about the smashed handling-machine I came upon the red weed again, and found the Regent's Canal a spongy mass of dark-red vegetation.'

Materialising first in London's Hyde Park to house 'The Great Exhibition of the Works of Industry of All the Nations' in 1851, then replicated with an additional glass dome two years later at the 1853 New York World's Fair, the Crystal Palace continues to reappear in locations as diverse as Madrid, Sydenham Hill and the Philippines throughout the latter half of the nineteenth century. A huge structure of glass and ironwork, it can never be anything other than temporary. At the same time its arching and exposed framework, suggesting skeletal remains rather than the ruins left behind by traditional stonework, offers the illusion of permanence – even as the shattered walls of Pompeii, along with the plaster casts of its long-dead occupants, are first exposed to public view. Whatever its location or condition, the Crystal Palace is an early meeting point for cinema and architecture, the glazed spans of Jugendstil buildings revealing their inner life – which is to say their inhabitants – as a form of unconscious content. Solid walls designed to keep out the elements also keep out information; seen as one vast sequential ordering of movie frames in space, the Crystal Palace brings representation and allegory together as an extended narrative of exotic machines, exotic plants and exotic people. Its apotheosis is the Winter Garden built in 1870 at the command of King Ludwig II of Bavaria: located on the roof of the Residenz palace in Munich, it not only contains tropical plants and insects that can be enjoyed all year round but also an ornamental indoor lake, a grotto, painted murals, pavilions and an Indian royal tent, plus a range of controllable lighting effects. 'The King invited me to take a centre seat

36, 53
35, 51

at the table and gently rang a little hand-bell,' one royal visitor recalls. 'Suddenly a rainbow appeared.'

Created in Paris at the height of the *Belle Époque* to commemorate the end of the Terror, Lobster Thermidor takes its name from the month in the French Revolutionary Calendar when Robespierre and Saint-Just both went to the guillotine. Involving the elaborate stuffing of an emptied lobster's shell, the dish suggests just how far Nerval's favourite pet has come since their mid-century strolls together around the Royal Palais Gardens. Where once live lobsters inhabited glass tanks in urban restaurants, awaiting selection by discerning customers, the domestic aquarium becomes a standard feature in the nineteenth-century home. Harking back to older, more mythical understandings of the sea as a source of dreams, the mermaid, the sea nymph and the lobster serve as reminders of a lost world that the industrial revolution has helped to create. The scales of the mermaid, her association with the ocean's foaming waves, together with the elegantly sculpted shell of the lobster, can be found in the Art Nouveau or Jugendstil line. Embodying the hardening and hollowing out of form in the late nineteenth century, such alien presences haunt the artificial worlds of the domestic fish tank and herbarium. Defined in terms of the technology that sustains it within the urban environment, the exotic quickly transforms high culture into a pathological condition. Even as the art critic Walter Pater speaks of the aesthetic intelligence's 'intense diamond hard flame', his image is derived from the plumber's blowtorch.

25, 83

By the century's end plumbers and engineers even have the power to dethrone kings. In April 1886, Ludwig II's lavish spending on exotic architectural fantasies, such as the castle at Neuschwanstein which is dedicated to the operas of Richard Wagner, is brought to an abrupt close when the company supplying his castles with water and gas sues for non-payment of bills, thereby opening up the floodgates to other creditors seeking payment. Within two months of this civil action, Ludwig is deposed as mentally unfit to rule and is later found drowned in Lake Starnberg, outside Munich. As a grim postscript to his tragic death, King Ludwig's Winter Garden is dismantled in 1897 after it is

31, 52

discovered that water from the ornamental lake has been leaking onto the floors below.

· 42 ·

'Take a Break – Have a Lobster'
  – Sign outside a seafood restaurant, Vienna

Perverse and ideal, lifeless yet rhythmic, the Jugendstil double curve freezes architecture into strange new forms, prompting Walter Benjamin to characterise it as 'the dream that one has come awake'. Caught between Brooklyn and the Atlantic Ocean, Coney Island is by the start of the twentieth century a wildly hallucinatory expression of that dream. 'It is marvellous what you can do in the way of arousing human emotions by the use you can make architecturally of simple lines,' declares Frederic Thompson in 1903. Decorated with over a million electric lights, his Luna Park invites its visitors to step onto the surface of the Moon in search of pleasures 'not of this earth', making it one of Coney's most successful institutions. Its main rival is Senator H Reynolds's Dreamland, whose main attractions include a submarine ride, 'Inhabitants of the Deep', 'the End of the World' and 'the Fall of Pompeii'. Access to these delights is granted by passing beneath the vast plaster sailing ships at its entrance, suggesting that only by submerging themselves beneath the waves can the public reach Dreamland. Dismissed by socialist author Maxim Gorky as a 'cheap, hastily constructed toyhouse for the amusement of children', Coney Island combines magic and engineering to transform itself into an Atlantis Risen for the masses. This is where the public now comes to have its intelligence underestimated. Writing in August 1905, campaigning journalist Lindsay Denison denounces Coney Island as the 'concentrated sublimation of all the mean, petty, degrading swindles which depraved ingenuity has ever devised to prey upon humanity'. At the height of its busiest season, the police come by regularly at dawn to clear the corpses off the beach.

35, 51

In 1911, set ablaze by a stray electric spark, Dreamland takes just three hours to burn to the ground. Three years later Luna Park also goes up in flames. By then Frederic Thompson has turned his attention to the Hippodrome, the vast theatre complex he has established on New York's Sixth Avenue, boasting 'a huge stage with mechanical means of depressing sections of it and a tank beneath, into which lines of show-girls could march and disappear'. This effect is originally created for *A Yankee Circus on Mars:* a theatrical revue dating from 1906 which tells of a stranded theatrical troupe invited to 'remain permanently and to become inhabitants of that far-away planet'. The show's climax is an amazing piece of aquatic choreography in which sixty-four 'diving girls' descend a staircase in squads of eight into the Hippodrome's vast arti-ficial lake; they 'walk into the water until their heads are out of sight', an effect so moving that 'men sit in the front row, night after night, weep-ing silently.' On Easter Sunday 1917, the Hippodrome is the chosen venue for a performance of Berlioz's *Requiem* – conducted by Edgard Varèse, making his US debut – as a memorial to those lost in the sink-ing of the SS *Vigilante* by a German submarine. From outer space to the ocean's depths, from a display of drowning to a commemoration of those drowned in a hostile act, the staging of Berlioz's *Requiem* at the Hippodrome conjures up images of immersion. The performance itself involves over 150 musicians and a choir of 300. Berlioz had originally conceived of placing four fanfares at the cardinal points of the church of Saint-Louis des Invalides, where the work was first performed in 1837, plunging his audience deep into the sound of its antiphonal brass. Prior to the Hippodrome performance, the Scranton Oratorio Society holds rehearsals for the male chorus, most of whom are Philadelphia miners, at the bottom of a mineshaft into which a piano had to be lowered for the purpose. Five days later United States Congress will vote to go to war with Germany.

20, 55
62

That same year US Navy recruit Buckminster Fuller examines water bubbles forming in the wake of his ship outside Brooklyn's Naval Yards and wonders how they manage to form perfect spheres without calculating pi. 'All true insight forms an eddy,' notes Walter Benjamin. 'To swim in time against the direction of the swirling stream.' In other

52

words, to yearn for the sea is essentially regressive: a confluence of womb, vulva and wave in which memory and oblivion are associated with the eternal flowing of water. As early as 1907 Georges Méliès films a silent adaptation of Jules Verne's *Twenty Thousand Leagues under the Sea* featuring aquatic encounters with half-naked sea nymphs. In 1917 a live-action version of the novel is filmed off the Bahamas using the 'photosphere', an underwater chamber allowing a camera crew to capture actual undersea shots of the submerged *Nautilus*. In the meantime John Milton Cage, inventor and father of the American avant-garde composer, successfully tests a prototype submarine capable of staying submerged for twenty-four hours straight, earning him a telegram of congratulation from President Woodrow Wilson.

59

20, 54

· 43 ·

And even in Atlantis of the legend
The night the seas rushed in,
The drowning men still bellowed for their slaves.
      – Bertolt Brecht, 'A Worker Reads History' (1928)

24, 76

The flooded antediluvian structures of Atlantis seem both as empty and as full as the transparent skyscrapers of tomorrow. When Fritz Lang first conceives of his silent classic *Metropolis*, detailing life in the city of the twenty-first century, it is the glittering towers of New York he has in mind. Extending itself vertically above and below ground, with no suburban sprawl to absorb the forces of progress and social unrest, no wonder the greatest threat to Metropolis is the deliberate flooding of its lower depths, where the workers' homes are allowed to disappear beneath the waves. 'That vertical city of the future we know now is, to put it mildly, highly improbable,' objects H G Wells in his review of Lang's film. 'Even in New York and Chicago, where the pressure on the central sites is exceptionally great, it is only the central office and entertainment region that soars and excavates.' Even so, visionary architect Hugh Ferriss starts sketching out the imposing

36, 53

vertical labyrinths which he confidently predicts will overwhelm the American skyline, 'In the future,' Ferriss claims, 'with the evolution of the cities, New Yorkers will literally live in the skies.' Unfortunately, the colossal bunkers, looming towers and skyscraper bridges which he devises in one heavily-shaded rendition after another only serve to reduce the heavens to a tiny grey blur, neither empty nor full, glimpsed from far below.

29, 55   Following on from Edison's defeat of the Martians, publisher Hugo Gernsback includes science-fiction tales in such technical journals as *Modern Electronics* (founded 1908), *The Electrical Experimenter* (1913) and *Science and Invention* (1920). In 1926, just as *Metropolis* is released to the general public, he launches *Amazing Stories*, which contains nothing but 'scientifiction' as Gernsback chooses to describe such material. It is courtesy of *Amazing Stories* that Buck Rogers has his earliest adventures in the twenty-fifth century. A syndicated comic strip starring Buck Roger appears in 1929: the first to feature rocket ships, atomic blasters and journeys to other planets. That same year Hugh Ferris publishes *The Metropolis of Tomorrow*, his grand design for life in the distant
31, 52   world of 1975. However, the Stock Market Crash of 1929 also means that his 'Visions of the Titan City' will never be realised. Skyscrapers are now for jumping off, not completing. The grand monoliths of tomorrow are incorporated instead into such design fantasies as the Twentieth Century Fox studio logo. Where Ferriss can only overwhelm with his impressionistic charcoal renderings of the future, pulp magazines like *Amazing Stories* thrill their readers with hard-edged, square-jawed depictions of powerful machines, shapely heroines and bolt-headed robots. In the work of popular cover artist Frank R Paul, every fantastic detail seems to float against broad planes of light, line and colour. The overall effect is a little like comparing a Gothic cathedral with a pinball table.

· 44 ·

Surrealism is irrezistieble and terifai-ingli
conteichios
Bieur! Ai bring ou Surrealism
*   – Salvador Dali, New York, 1936*

Meanwhile the Great Depression throws up some unlikely new heroes, such as Alfred W Lawson whose Direct Credits Society wins an
extraordinary mandate from the huddled masses. A professional
baseball player, aviator and the inventor of the first commercial airliner as well as a 'trans-oceanic float system', Lawson is less a chieftain of
industry than one of its more inspired berserkers, denouncing modern
mathematics as 'a cheat's invention used by people to defraud one
another'. As economic theory, Direct Credit may be worthless, but it is
nothing compared with 'Lawsonomy', a detailed and highly complex
body of knowledge which will, its creator boasts, explain 'everything
about everything to everybody'. In such self-published books as
Manlife and *Creation*, Lawson outlines a cosmology based upon
the universal forces of 'Suction' and 'Pressure', which finds a curious
echo in the theories of Viennese engineer Hanns Hörbiger. Declaring
objective science to be a 'pernicious invention' and mathematics
'nothing but lies', Hörbiger propounds a Doctrine of Eternal Ice in
which the universe responds solely to the forces of 'Attraction' and
'Repulsion'. A series of ice moons have thereby been irresistibly pulled
down from the sky to the Earth, he argues, drastically altering the
planet's gravitational field and producing huge tidal waves, one of
which caused the watery destruction of Atlantis. Originally published
in 1913, Hörbiger's *Glacial Cosmology* finds a new and enthusiastic
coterie of readers among theorists eager to rid the scientific universe
of 'Jewish physics', which Professor Philipp Lenard of Heidelberg
identifies as 'a phantom and a phenomenon of degeneration of   32, 66
fundamental German physics'. Artists, scientists and intellectuals
threatened by the Nazi rise to power start the long journey across the
Atlantic, seeking refuge in the United States.

Throughout the 1920s and 1930s Edgar Cayce, the 'Sleeping Prophet' of Virginia Beach, describes in great detail daily life in ancient Atlantis. Going into a deep trance, lulled by the sound of the Atlantic waves, he warns that the poles of the planet will one day be violently reversed, allowing the sunken towers of Atlantis to rise again. American news-stands are now heaving under an increasing number of science-fiction magazines such as *Air Wonder Stories, Science Wonder Quarterly, Astounding Stories, Startling Stories,* and *Marvel Science Stories.* Can the amazed, the astounded and the startled keep pace with progress? 'It is scarcely an exaggeration to say that any contemporary conscious-ness that has not appropriated the American experience, even if in opposition, has something reactionary about it,' Theodore Adorno remarks of his experience writing *Dialectic of Enlightenment* in Los Angeles with fellow exile Max Horkheimer. In 1931 a 'Fête Moderne: A Fantasie in Flame and Silver' is held at New York's Hotel Astor to celebrate the future; among the skyscraper replicas, science-fiction costumes and space-age decor, one woman comes dressed sim-ply as the 'Basin Girl', complete with faucets, pipes and mirror. The bathroom has replaced the fish tank in the modern household as its principal watery medium. The plumber, the architectural designer and the structural engineer have become the real heroes of the future. At the 1933 Chicago World's Fair, while the United States climbs out of the Depression, Buckminster Fuller demonstrates his new Dymaxion car, and the first Buck Rogers film is premiered: a ten-minute epic entitled *An Interplanetary Battle with the Tiger Men of Mars.*

27, 60   It is, however, an erotic dream of lions that troubles Salvador Dali's sleep on his first night in New York. Only when their roaring continues into the Surrealist painter's waking life does he discover that his hotel bedroom looks out over the city's zoo. With a precision lacking in the darkened abstract shadings of Hugh Ferriss, Dali's 'paranoid-critical' method responds magnificently to New York, where the perspectives offered by its crowded skyline establish increasingly ambiguous relationships

10, 72   between figure and ground. 'NEW YORK DREAM – MAN FINDS LOB-STER IN PLACE OF PHONE' runs the caption to a drawing executed by Salvador Dali for the illustrated magazine *American Weekly* in 1936.

From now until the end of the 1930s Dali's travel plans will take him back   6, 84
and forth across the Atlantic, commuting between Europe and Manhattan by passenger liner only, as the great painter is terrified of flying.

· 45 ·

By satisfying each new desire it arouses, progress sells itself to the
masses. But progress also has its discontents. There are those who   28, 57
quickly grow dissatisfied with such thrusting materialism. The masses
will always be in search of something more. In 1924 Colonel Percy H
Fawcett disappears into the jungles of South Western Brazil in search
of evidence for the historical existence of Atlantis and is never seen
again. That same year Lewis Spence publishes *The Problem of Atlantis*,
in which he argues that the legendary lost island cannot fit in with the
physical facts of adaptive evolution without some gigantic catastrophe
having occurred to submerge it forever within human consciousness.
Perhaps a giant meteor struck the Earth or a shift in planetary alignment
took place, producing huge floods and tidal waves. Behind this proposition is the idea that celestial bodies are capable of influencing the
world, drastically affecting its landmasses and their inhabitants, throwing everything into chaos, and that humanity somehow welcomes it.
From his couch by the Atlantic ocean Edgar Cayce announces, deep in
a self-induced trance, that 'man brought in the destructive forces that
combined with the natural resources of the gasses and electrical forces,
that made the first of the eruptions that awoke from the depths of the
slow-cooling earth.'

Joining Fritz Lang in self-imposed exile from the Nazis is a growing
Hollywood colony of German writers and directors. Among them is   33, 62
Curt Siodmak, who scripted *FP1 Does Not Answer,* a futuristic melo-
drama set aboard a gigantic floating platform similar to the one invented by Alfred Lawson, and *The Tunnel,* about a subterranean motorway
constructed under the ocean to link Europe with America. Works of
engineering are myths of transition: both movies are transatlantic in

location, feature engineers, aviators and industrialists as their heroes and look towards America for their future. By 1937, Siodmak is in England scripting a futuristic murder mystery, *Non-Stop New York,* set almost entirely aboard a large transatlantic airliner. As both title and story-line suggest, Siodmak remains intent upon following his engineer heroes on their long trajectory towards America and the future. Meanwhile Atlantis has become one of the regular stops along the way to an uncertain tomorrow. As a sequel to *The Problem of Atlantis*, Lewis Spence publishes *Will Europe Follow Atlantis?* Written during the Nazi rise to power, it argues that the moral excesses of the modern world may well lead to a second deluge. By now even Buck Rogers has ventured down to the watery ruins of Atlantis.

Hermann Rauschning, the ex-Nazi leader for Danzig and former confidant to Adolf Hitler flees Germany for America, bearing strange tales of what goes on inside the Führer's brain. Between 1939 and 1941 he publishes a series of astonishing revelations. 'It is impossible to understand Hitler's political plans,' Rauschning cautions his readers, 'unless one is familiar with his basic beliefs and his conviction that there is a magic relationship between Man and the Universe.' Siodmak's engineer heroes continue to be dismissed in favour of the most apocalyptic belief systems. 'We are often abused for being the enemies of the mind and spirit,' Hitler likes to boast. 'Well, that is what we are, but in a far deeper sense than bourgeois science in its idiotic pride could ever imagine.' Scientists and physicists are now signing up to Hanns Hörbiger's Doctrine of Eternal Ice. Among them are Hermann Oberth and Willy Ley, confounders of Germany's Society for Space Travel whose membership includes a young Wernher von Braun. 'Our revolution,' Hitler asserts, 'is a new stage, or, rather, the final stage in an evolution which will end by abolishing history.' Or as Herr Rauschning puts it: 'At bottom every German has one foot in Atlantis.'

108

• 46 •

To make in ourselves a new consciousness, an erotic sense of reality, is to become conscious of symbolism. Symbolism is mind making connections (correspondences) rather than distinctions (separations). Symbolism makes conscious interconnections and unions that were unconscious and repressed. Freud says symbolism is on the track of a former identity, a lost unity: the lost continent, Atlantis…   27, 60
    – Norman O Brown, Love's Body

Throughout the twentieth century, the masses are treated as an imperfect medium for the realisation of ideas. The industrial exhibition consequently becomes a machine designed for the historical processing of crowds, giving the public direct access to such concepts in   26, 71
modern living as futuristic architecture, high-speed trains, prefabricated homes, multilane highways and robots. Taking place outside Manhattan at Flushing Meadows almost a full century after its 1853 predecessor, the 1939 New York World's Fair, has 'Building the World of Tomorrow' as its theme. Visitors to the Westinghouse 'Hall of Electric Living' are invited to marvel at 'Elektro the Moto-man' as he performs mystifying feats of deduction or thrill to the 'Battle of the Centuries' between a frowsy-looking housewife and an automatic dishwasher. Elsewhere General Motors' Futurama pavilion presents life in year 1960, while RCA proudly displays its first television set. Located outside   14, 51
the beaux-art symmetry of the main attractions, impassively arranged around the abstract forms of Wallace Harrison's Trylon and Perisphere,   11
is the fair's Amusement Zone. Notable among its many sideshows offering 'streamlined festivity for the World of Today and the World of Tomorrow' is 'Dali's Dream of Venus': a combination dime museum, fun house and Surrealist art show that represents, so its instigators claim, 'an attempt to utilize scientifically the mechanisms of inspiration and imagination'. Even so, the fair's original contract, drawn up with Dali's New York dealer Julien Levy and architect Ian Woodner, calls for a work entitled 'Bottoms of the Sea'. A separate business deal with a US manufacturer of a new kind of soft rubber that keeps its shape underwater helps to establish the exhibit's theme, while a newspaper

report claims Dali's real interest to be the 'bottom of men's minds and everyone knew that there was no end to them'.

Unlike the Perisphere, inside which a scale model of 'Democracity: the Metropolis of the Machine Age' is on display, 'Dali's Lovely Liquid Ladies' are housed within a lopsided structure of distended coral formations and marine accretions set in white plaster. Spines and teeth are mixed with splayed legs and bared breasts beneath a giant reproduction of Botticelli's Venus rising serenely from the sea – the fair organisers having firmly rejected Dali's idea of replacing her face with a monstrous fish head. Inside is Aphrodite's grotto as mythological labyrinth, where a half-naked Venus reclines upon a seemingly end-less bed, surrounded by lobsters and bottles of champagne. In a deep sleep, her dream takes an underwater form, which is what the public has really come to see. 'A long glass tank,' *Time* reports, 'is filled with such subaqueous decor as a fireplace, typewriters with fungus-like rub-ber keys, rubber telephones, a man made of rubber ping-pong bats, a mummified cow, a supine rubber woman painted to resemble the keyboard of a piano. Into the tank they plunged living girls, nude to the waist and wearing little Gay Nineties girdles and fishnet stockings.' The enrobing of 'Dali's Nude Aquarium' represents a complex series of mythological transformations. The birth of Aphrodite, emerging unclothed from the bloodied water after her father's castration, al-ready has a twentieth-century incarnation at the World's Fair. DuPont's Wonder World of Chemistry has a 'Princess of Plastics' emerging from a giant test tube dressed entirely in nylon: 'lace evening gown, stockings, satin slippers and undergarments', a press release reveals. Meanwhile publicity shots for 'Venus's Prenatal Chateau beneath the Water' show Dali tricking out naked women in such undersea finery as raw oysters, eels and mussels. One model sports a lobster in her hair with another lain across her lap; a second has a lobster curled over her mound of Venus, evoking Jugendstil nightmares. 'Like the lobster of the deep, I am dressed in my skeleton,' the goddess intones on the 78 rpm gramophone record released to promote the Dream of Venus show. 'Within is the rose-coloured flesh of dreams, more gentle than honey.'

• 47 •

'Come on Blitzcreak!' Evonne Kummer, one of the darkly beautiful charmers in 'American Jubilee' of the World's Fair 1940 gives her prize lobster a breezing in preparation for the Crustacean Stakes Classic to be run around the Perisphere Pool next Friday, May 31st. At least fifteen hardshell speedsters are expected to take part in the contest, promoted by Stanley McGinnis. Blitzcreak surprised even his most optimistic backers with the phenomenal speed of 20 feet an hour in his workout with Evonne.
　　　– Director of Publicity, New York World's Fair

The old-fashioned corsets, garters and fishnets worn by the women as they swim about the glass tank also evoke memories of the 'Dive Show', an aquatic form of burlesque featuring live mermaids and Edwardian bathing belles. The organised human mechanics of the corps de ballet in Act II of *Swan Lake* have been transformed into the floating aquatic dream evoked by the underwater sporting of half-naked sea nymphs inserted by Méliès into his film adaptation of Jules Verne's *Twenty Thousand Leagues under the Sea*. It's an old trick, but it still plays well to the crowds on the midway. While planning their Surrealist side-show, Levy and Woodner are offered some free advice from Billy Rose, creator of the Aquacade, a synchronised swimming gala staged at the World Fair's Fountain Lagoon. 'Anything writ in water will succeed,' the old showman confides, 'lagoons fountains, aquacades, ice coolers anything you please but the public is disposed towards water.' In 1940, as Germany continues its armed assault on the rest of Europe, the publicity department of the New York World's Fair issues a press photo-graph of brunette Evonne Kummer taking 'Blitzcreak', her aptly-named pet lobster, for a walk around the Perisphere – in imitation of Nerval, she leads the gentle crustacean around on a length of ribbon.   118

It looks as if someone has finally lifted the receiver on Dali's lobster tele-phone. Transformed through the medium of water, Hörbiger's doctrine of Eternal Ice and Lawson's principles of Suction and Pressure stand revealed as the painter's 'paranoid-critical' method played out on a cosmological level. 'Here is what we can still love,' Dali writes approvingly of Art Nouveau architecture, 'the imposing block of those rapturous and frigid structures scattered across all of Europe.' Images ripple and bend within the Dive Show's glass tank, rendering it empty and full at the same time – imitating 'the world of the hardened undulations of sculpted water' he claims to have discovered in the Jugendstil line. Like some ancient dream, a swirling assemblage of cultural flotsam and jetsam is brought erotically to life by the presence of the half-naked female divers, 'seen at close range and a trifle water-magnified', according to *Time*. As if to validate the introduction of indoor plumbing to the modern home, Dali's chorines swim in pure New York tap water. 'This isn't water,' complains one group of visiting tourists, thinking they have seen through this particular trick, 'it's compressed air.'

20, 87

Except that the inside of the tank is murky on opening night. Dye from the bathing belles' gloves dirties the water of the unconscious mind; plus, as a backdrop to the girls' gyrations, Dali has painted Vesuvius erupting burying the ruins of Pompeii under a cloud of volcanic ash. Dali himself has fled his own dream and is already back across the Atlantic, working with the Ballets Russes on the staging of a ballet inspired by King Ludwig II, provisionally entitled *Venusberg*. Very soon the world reflected in the arrangement of pavilions at Flushing Meadows will no longer exist. In May 1939 US Navy submarine *Squalus* sinks off the New Hampshire coastline. Having been denied an official presence at the World's Fair, Germany demands the Czechoslovakian exhibit be destroyed after the Nazi invasion of that country. A public subscription keeps it open until the fair finally closes in 1940. By then 'Dali's Dream of Venus' has been renamed '20,000 Legs Under the Sea'; and Walter Benjamin has taken his own life by swallowing morphine while attempting to reach America on papers arranged for him by Max Horkheimer.

• 48 •

'I have seen the future'
  – Lapel pin sold at 1939 World's Fair

Condemned to disappear forever, it is the fate of all industrial fairs to be looked back on; as Walter Benjamin observes, they busy themselves with preparations for the past, shaping the history of our future. A post-card issued for the 1939 World's Fair links the 1853 Crystal Palace 'across the fairway of time' with the Trylon and Perisphere, custodians of the new model 'Democracity', showing 'how we should be living today'. While it still stands, the fair's motto remains unequivocal: 'Science Finds, Industry Applies, Man Conforms.' Dali's cover illustration for the catalogue to his 1939 show at the Julien Levy Gallery in New York shows the Trylon and Perisphere as distorted ruins of their former selves against a black sky. King Ludwig II of Bavaria, Wagner's protector and patron, left instructions for all his fairytale castles to be blown up after his death – including the famous Venus Grotto at Linderhof, supported by an ingenious framework of iron girders, with its cement stalagmites, 'artificial waterfall and wave-making machine'. The twilight of the gods, may end in fire, as Wagner's music drama dictates, but it starts with a devastating flood as the Rhine bursts its banks. Floods are always biblical: apocalyptic. In April 1945 Hitler and Goebbels commit suicide in their command bunker beneath the ravaged streets of Berlin. In one final gesture of defiance, Joseph Goebbels orders the Berlin subway system to be flooded as punishment for the hundreds of families who have sought refuge there. Like some reinvented scene from *Metropolis* or *The Tunnel*, Atlantis has finally fallen.

86

29, 55

92

By April 1945, six years after the New York World's Fair first opened to the public, little remains of the engineering marvels that once dominated Flushing Meadows. The towering steel curves of 'The Parachute Drop' have been dismantled and shipped over to Coney Island, where the ride now stands outlined against the Atlantic drifts, alongside such attractions as the Wonder Wheel and the Cyclone. Reaching across the fairway of time in the July 1945 issue of *The Atlantic*, Vannevar

54

Bush revisits the architecture of the Crystal Palace but in a powerfully miniaturised form. 'A spider web of metal, sealed in a thin glass container,' the director of wartime scientific research in the US enthuses, 'a wire heated to brilliant glow, in short, the thermionic tube of radio sets, is made by the hundred million, tossed about in packages, plugged into sockets – and it works!' Recalling the 1939 World's Fair in this groundbreaking essay, simply titled 'As We May Think', Bush sketches out a future in which the world has 'arrived at an age of cheap complex devices of great reliability; and something is bound to come of it.' Among the examples he cites are the typewriter and telephone, both of which form part of the underwater décor in Dali's Nude Aquarium. Memories of the half-naked chorines dreamily handling these rubberised devices seem to have eroticised Bush's descriptions of technology in action. 'A girl stroked its keys,' he recalls of one device displayed in 1939, 'and it emitted recognizable speech.' The same image reappears in his unsettling description of a stenotype operator: 'A girl strokes its keys languidly and looks about the room and sometimes at the speaker with a disquieting gaze.' Bush may also recall that an ambitious mural at the Communication Building depicted 'the progress of the art of speech from the age of myth to the age when myths were household fixtures'.

Modernism for the masses requires new myths: ones that will distance them from those of the past. Caught between the old and the eternal, the modern represents only a moment of temporary respite. Also in the summer of 1945, having introduced readers of *Amazing Stories* to 'Mantong' – an original dialect of Atlantis incomprehensible to existing language experts who 'operate on incorrect principles' – former Detroit welder Richard Shaver describes his former life in 'Sub-Atlan', a city located miles below the fabled lost continent. In this earlier incarnation, Shaver reveals, he was a descendant of ancient, god-like creatures who had come from outer space 150,000 years ago to colonise Atlantis and Lemuria. Both races soon discovered that the sun was beginning to emit poisonous radiation which ravaged their bodies and stripped them of their immortality. Some returned to space in search of another Atlantis to colonise, while others burrowed deep into the bowels of the Earth,

76

26, 60

creating vast underground cities accessible only through caves located on the planet's surface. Published in *Amazing Stories* at the end of World War Two as a true account of actual events, Shaver's tales of Sub-Atlan and its terrible legacy prepare the American public for the first reported sightings of flying saucers in the spring and summer of 1947.

· 49 ·

'Enter the shell of my house and you will see my dreams.'
   – 'The Dream of Venus' souvenir record

Immanuel Velikovsky, a former student of Freud's pupil Wilhelm Stekel, visits the USA on a temporary visa in 1939 just as war breaks out in Europe. He subsequently becomes a resident of New York City, where he studies Hanns Hörbiger's Doctrine of Eternal Ice in the public library before developing a theory of his own. Published as an educational handbook in 1950, *Worlds in Collision* argues that the planet Venus was pulled out of Jupiter's enormous mass by a passing comet, which went on to drag Mars into its present orbit and visit all manner of cataclysmic events on the Earth. Born out of water fertilised by divine blood, Venus's dream is another deluge. The floods and tidal waves that have swept through our collective memory are, Velikovsky suggests, but psychic recurrences of the same cosmic incident. Buffeted by unimaginable forces and planetary alignments, the human image stands exposed in the middle of the twentieth century as a twisting column of water: a dynamic and complex form constantly shifting in space.

Psychoanalysis plays an integral role in the reinterpretation of human history. Rising naked from the sea inseminated by her father's bloodied genitals after being cut from his body by his own son, Venus presides over a world drowned in amniotic fluid. Swirling in these erotically charged currents, the 'typewriters with fungus-like rubber keys' and 'rubber telephones' Dali sets among his half-naked chorines convey the authentic whisperings of the deep: messages from beyond, relayed without the aid of wires by virtue of the imagination. 'I cannot help mentioning how often mythological themes find their explanation through dream interpretation,' Freud observes. 'The story of the Labyrinth, for example, is found to be a representation of anal birth; the torturous paths are the bowels, and the thread of Ariadne is the umbilical cord.' Somewhere near the back of Dali's Nude Aquarium, the Minotaur is reborn as a 'mummified cow' placidly reclining against a backdrop of Vesuvius erupting. It was only with the aid of a special harness engineered by Daedalus to resemble a cow that the wife of King Minos could couple with a white bull and thereby give birth to the monster that haunts the Labyrinth. The bull was sent by Neptune and her love for it came as a punishment from Venus. Neptune founded the Atlantean race by fathering ten children with a mortal woman whom he surrounded with canals. Mars and Venus are born out of a single interplanetary upheaval. *Worlds in Collision,* its status as a textbook hotly contested by the scientific establishment but read avidly by the general public, offers a retelling of the old industrial myths in the name of a new cosmology. Both Atlantis and the Martians appear increasingly transitional in nature. How will they survive in the great black emptiness of space?

· 50 ·

Among the stylistic elements that enter into jugendstil from the iron construction and technical design, one of the most important is the predominance of... the empty over the full.
    – Walter Benjamin, The Arcades Project

From Atlantis to Mars: transition is a means of deferring an unresolved conflict. Planting a foot firmly in the World of Tomorrow, the Westing-house Corporation buries a time capsule containing memorabilia of the 1939 World's Fair beneath Flushing Meadows. Its contents are not to be unearthed until the year 6939: five thousand years into the future. The

5, 61

5, 90

company deposits a second capsule, ten feet away from the first, at the close of the 1964 New York World's Fair, which is also held in Flushing Meadows. This one is also scheduled to be opened in 6939. How much time does change require to take effect? Atlantis, after all, disappeared in a single night. Westinghouse is not alone in repeating itself so optimistically. General Motors reinvents its Futurama pavilion from the previous New York World's Fair, now depicting the world of 2024, while Walt Disney offers the public a revolving 'Carousel of Progress' which, as its name implies, also requires constant updating. Emblematic of this encroachment upon the present is the Unisphere, a hollow steel representation of the Earth as seen from space, its oceans little more than empty outlines upon its surface. Built on the foundations of the 1939 Perisphere, it stands in the middle of a large reflecting pool, surrounded by a ring of fountains designed to draw attention away from the pedestal supporting it. Not surprisingly, Robert Moses, the man in charge of the 1964 World's Fair, is also responsible for placing an aquarium over Dreamland's charred remains on Coney Island – less surprisingly still, David Woodner, who designed the outer shell to 'Dali's Dream of Venus', once worked for him.

The unsupported sphere floating in empty space is part of the new post-war cosmology. Dedicated to 'Man's Achievements on a Shrinking Globe in an Expanding Universe', the Unisphere looks outwards where the Perisphere turned itself in upon the model city displayed at its centre. By 1964 the Soviet cosmonaut and the US astronaut have 19, 110 come to represent the masses on a planet whose surface has been mapped and vectored in its entirety. The Russian Sputnik programme has seen to that; while USS *Nautilus*, powered by Westinghouse nuclear engines, sails under the polar icecap – but only after its namesake from Verne's *Twenty Thousand Leagues under the Sea* becomes part of a walkthrough attraction at Disneyland. Like the other submarines in America's nuclear fleet, the *Nautilus* has a full set of Martin Denny tapes in its record library. Providing an immersive environment in a hollow world, the exotic easy-listening experience lends itself readily to underwater themes: releases from the period include Denny's *The Enchanted Sea*, Nelson Riddle's *Sea of Dreams*, Les

Baxter's *Jewels of the Sea* and Alexander Lazlo's *Atlantis in Hi-fi*, featuring tracks such as 'Traffic Of A Sunken City', 'Conference Of The Sea Gods', 'Rapture Of The Deep' and 'Exiles Of Atlantis'.

To confront the unconscious as an environmental fact is to encounter a hollow world. When water is a medium, information becomes another element. The best sculptures in America remain its drinking fountains and bathroom fixtures. 'We have all means to convey sounds in trunks and pipes, in strange lines and distances,' boasts one of the islanders in Francis Bacon's *The New Atlantis*. The Utopian relationship between music and plumbing is given modern expression on such albums as *Music for Bathroom Baritones and Bathing Beauties* and *American Standard's Music for Bathrooms,* offering the soundtrack to a new, intimately self-contained existence. Even before the fountains are switched on in Flushing Meadows for the 1964 World's Fair, *Archigram 3* goes into circulation. Partially inspired by the theories of Buckminster Fuller, the magazine's content is categorised under three separate headings: 'Bathrooms', 'Bubbles' and 'Systems'. These will form the architecture of the Space Age. Mars awaits us.

Duration,
Or the Birth of Chance
Out of the Spirit of Music

'Does music perhaps belong to a culture in which the dominion of men of power, of every kind, has already come to an end?'

Friedrich Nietzsche
Unpublished Fragment from 1880

· 51 ·

Perhaps, Kretzschmar said, it was music's deepest desire not to be heard at all, not even seen, not even felt but if that were possible, to be perceived and viewed in some intellectually pure fashion, in some realm beyond the senses, beyond the heart even. Except, being bound to the world of the senses, it must again strive for its most intense, indeed sensual realization, like a Kundry who does not wish to do what she does and yet flings soft arms around the neck of the fool.

— *Thomas Mann*, Doctor Faustus

When you come right down to it, the future is just people laughing together in a room. From Monsanto's House of the Future, opened to the public in Disneyland the same year Russia launched its first Sputnik, to the carefully regulated data streams of the Microsoft Home, modelled upon Bill Gates's own Seattle mansion, this is the image most often projected onto tomorrow, even when we are not permitted to share in the joke. People laughing together in a room, at ease with one another in a private space: it represents the most basic design for modern living, holding out to us the prospect of comfort and security, coupled with a feeling of weightless belonging in a world of bright promise. Utopia is both the domination of nature taken to extremes and an unflinching inability to confront the everyday world in any way. In structural terms, it can never be more than a temporary hoarding, where coming attractions are fly-posted on behalf of the

25, 74

future. The scraps and remnants overlap each other: a forgotten snatch of library music here, some frames from a promotional movie there, a moment at a trade fair, a picture spread in a glossy brochure or the flash of text from a television commercial. When the pieces fit together, we call it progress.

A 'Paris Street' on display at the 1900 Paris World Exhibition 'realizes, in an extreme manner, the idea of the private dwelling that is characteristic of *jugendstil*', Walter Benjamin remarks. The published account upon which he bases this observation does not describe an urban space so much as a series of staged entertainments. 'The satirical newspaper *Le Rire* has set up a Punch and Judy show,' its author notes. 'The originator of the serpentine dance, Loïe Fuller, has her theatre in the row. Not far away… a house that appears to be standing on its head, with its roof planted in the earth and its doorsills pointing skyward, and which is known as "The Tower of Wonders".' The private individual demands of an interior that it foster his illusions; the juxtaposition of puppets and dancers with such deliberately inverted wonders hints at the conceptual development of the modern home into a domestic environment. On closer inspection, their playful arrangement turns out to be equally illusory. These are not separate attractions stood up in a row to distract the visiting public but different versions of what is essentially the same house. The facades ranged haphazardly along this 'Paris Street' represent an architecture glimpsed exclusively in terms of the events it makes possible. Like the frames of an unfinished movie, sequentially imposed one upon another, they merely set the scene.

The staging of this particular illusion will take place without the benefit of music, however. The domestic interior is swiftly altered to suit the imposition of a pre-recorded soundtrack instead. Even in Utopia it turns out that music and indoor plumbing cannot coexist. 'We have also sound-houses, where we practise and demonstrate all sounds and their generation,' Bacon's islander reveals in *The New Atlantis*, carefully avoiding all mention of that which secretly does not wish to be heard. From bedroom to bathroom to lounge, the tension between public and private space has never been handled well by music, except as an

41, 97
42, 97

35

50, 88

expression of privilege or in a manner that renders solitude impossible. 'Satie invented Furniture Music as a means of getting into society (rented out for a party),' *Dadaphone* sneers in 1920, denouncing the composer of a 'utilitarian music' specifically intended not to be listened to. The change in title for this last issue of the periodical *Dada*, edited by Tristan Tzara, would seem to imply as much in itself.

Consider the impact of high-fidelity stereophonic sound upon the evolution of the suburban home from the 1950s to the present day. The positioning of loudspeakers, the acoustic reshaping of rooms, the introduction of modular shelving to accommodate tape decks and turntables, amplifiers and tuners, racks and cases: the modern home realises music's ultimate desire not to be heard by seducing the senses with its duplicate: sound reproduction. In the digital household of today it is still the sensual appeal of sound alone that compels Kundry to seduce the innocent Parsifal. The reproduction of sound, taken to the microscopic   3, 27
edge of the digital event horizon, is moving towards the tactile and the environmental, which is to say, towards the limitless background of our daily lives. The digitisation of sound takes music neatly out of earshot. The tactile instability expressed in the 'granular' play of zeros and ones suggests an equal instability of both medium and form. Music merges with texts, images, arguments and propositions. Incapable of resting long enough to resolve itself, such streamed data is never complete. It lacks definition in every sense of the word. To borrow an analogy from McLuhan, it is entirely possible that our society is no longer listening to   33, 60
music – we are watching television instead.   46, 64

• 52 •

'I won't create a pavilion,' the architect Le Corbusier instantly replies   76
when invited to design the Philips Pavilion for the 1958 Brussels   43, 63
World's Fair, 'I will create an "electronic poem" together with the vessel that will contain it. The vessel will be the pavilion, and there will be no façade to this vessel.' In other words, his modern 'Tower of Wonders'

is not to be turned upside-down but inside-out, presenting the visiting public with one large windowless interior for the three-dimensional projection of 'a work capable of profoundly affecting the human sensibility by audiovisual means… using the perhaps prodigious means offered by electronics: speed, number, colour, sound, noise, unlimited power.' Its conception marks one of the earliest instinctive understandings of how electronic media have surreptitiously transformed architecture into an event. Having been mechanically pulled apart during the previous century, the senses are to be recombined during the 1960s into a single delirious implosion of sound, light, colour and image. The container for this 'electronic poem' will be a lofty futuristic vessel created by the young Greek composer Iannis Xenakis; a featureless rhythmic series of peaks and parabolas executed in poured concrete. With no more 'façade' than the back of a movie screen, the Philips Pavilion is a twentieth-century sound-house, an inner space waiting to be filled – and Le Corbusier insists that Edgard Varèse is the only composer capable of doing so.

42

The Philips Pavilion officially opens on 17 April at Expo '58 then promptly closes again until 2 May, throwing Varèse into a rage. The complex sound system, designed to project amplified recordings via 425 individual loudspeakers into its starkly modernist interior, is not yet ready. 'Ever since the twenties an immense variety of Philips electro-acoustical equipment has reached its destination,' the Dutch electronics firm boasts in one its advertisements. 'Philips microphones, amplifiers, loudspeakers and sound recorders were the faithful witnesses of more than three decades of evolution in sound registration and reproduction.' A small portrait of Varèse accompanies the copy. Having briefly gone into the manufacture of electric organs, Philips focuses instead upon promoting stereophonic high-fidelity playback for the home. Music, so far as they are concerned, is no longer something to be played but reproduced. To this end, the company establishes the Philips Research Laboratories in Eindhoven: a studio for the creation of electronic music on reel-to-reel tape machines for release on the company's own record label. If nothing else, these experimental new works will be used to demonstrate the possibilities of experiencing stereophonic high-fidel-

17, 86

ity in your own living room. The transcription of sound onto magnetic tape offers Varèse the opportunity of creating works in which 'the composition itself is the orchestration'. For a composer who started out as a gifted conductor, the possibilities for controlling colour, intensity and velocity must have been self-evident. Both score and realisation at the same time, his 'electronic poem' will be its own interpretation. Even after seven months at the Philips Research Laboratories, however, his groundbreaking work will have to wait a few more weeks before the public can fully appreciate it.

What is waiting to be realised within the Philips Pavilion is a defining moment in the slow dispersal of music into the human environment. What began in the 'mystical abyss' shielding the orchestra from the audience's gaze at the Bayreuth Festspielhaus, thereby rendering Wagner's music an invisible participant in the drama unfolding onstage, now dissimulates itself further in an architecturally-organised interplay of sound and image. The electrically-generated sonic intensities created by Varèse in the studio at Eindhoven swell and recede through space accompanied by an array of abstract colour shapes projected directly onto the blank walls. 'Then pictures begin to appear on the wall,' notes *The New York Times*. 'Some are large and some small, and occasionally they join in huge images that seem to cover the asymmetric surfaces of the vaulting chamber... There are masks, skeletons, idols, girls clad and unclad, cities in normal appearance and then suddenly askew,' the report continues. 'There are mushroom explosions so familiar to newspaper readers and moviegoers in an era of atomic bombs. The score is not compounded of recognizable instruments. It is the work of a man who has been seeking for several decades to return music to a purity of sound that he does not believe possible in conventional music making.' *Poème électronique* is an early expression of the forthcoming delirium, a shock wave from some future space that only those already familiar with electronic media can register with any accuracy. 'Here,' according to the magazine *Radio et TV*, 'one no longer hears the sounds, one finds oneself literally in the heart of the sound source. One does not listen to the sound, one lives it.' Or, to be more precise, one lives in it.

31

41, 76

· 53 ·

We've had quite ENOUGH machine/machine/machine/machine
  – *El Lissitzky*

'One of the greatest assets that electronics has added to musical composition is that of metrical simultaneity,' Varèse declares in a public
lecture. 'My music being based on movements of unrelated sound
masses, I have long felt the need and anticipated the effects of having them move simultaneously at different speeds.' Simultaneity, the
most forceful celebration of the instantaneous, has preoccupied the
avant-garde since the beginning of the twentieth century – from the
Futurist *sintesi* based on the raw dynamism of the Italian variety theatre,
through the 'simultaneist' poems read aloud by multiple voices at the
Cabaret Voltaire in Zurich, to the embracing of circus ring and sideshow
by Cocteau and Les Six. In its entirety, Francesco Cangiullo's 'Detonation: Synthesis of All Modern Theatre' calls for one minute's silence to be
abruptly terminated by a single gunshot. Formal control of speed and
mass quickly becomes a defining principle; the organised spontanei-
ty of the modernist cabaret finds its precise counterpart in the factory

26, 66
114

assembly line. First published in 1911, just two years after Marinetti's
'The Founding and Manifesto of Futurism', Frederick Winslow Taylor's *Principles of Scientific Management* transforms the elimination
of waste and inefficiency from the workplace into a ruthless aesthetic. The political economies of time and motion devised by Taylor in
pursuit of standardised industrial production help to establish

40, 62

patterns of behaviour that make little distinction between humans and
machines. The production line is perceived as the coercive harnessing
of controlled spaces and routines integrating metal and flesh components. Taylorism, in short, is Reuleaux's kinematics expressed in social
terms.

41, 64

In the age of Eli Witney and Henry Ford, the industrialisation of culture
inevitably arises from the measurement of work and leisure. During the
early 1920s, after the mechanical inscription of sound is superseded
by the commercial recording of music, the composer George Antheil

claims that the traditional orchestra has become a thing of the past. Published in the pages of *De Stijl*, his 'Musico-Mechicano Manifesto' calls instead for 'vast music machines in every city' able to 'open a new dimension in man' by making the public vibrate psychically in time with their powerful rhythms. As no traditional instrument can ever hope to meet Antheil's demands, it is left to the remorseless whirring of the early silent-movie projector to synchronise people's responses to a single collective beat. In its cinematic depiction of the rhythmic connection between man, city and machine, *Metropolis* shows the workforce visually called to order by gigantic banks of factory sirens; their soundless swelling echoes Russian composer Arsenij Avraamov's revolutionary 'Symphony of the Factory Sirens', a synchronised performance staged across the city of Baku in 1922 and coordinated with the aid of flags and pistol shots. Antheil's own 'first major work' is scored for percussion and 'countless numbers of player pianos' to accompany Leger's 1924 silent movie *Ballet Mécanique*. The composer's rhythms reach across Leger's edited images; continuity is established through tempo and repetition. Antheil describes the piece as 'All percussive. Like machines. All efficiency. No LOVE. Written without sympathy. Written cold as an army operates. Revolutionary as nothing has been revolutionary.' Except that the mechanical player pianos, in their cold efficiency, refuse to stay in synch with each other, and human pianists are required for the realisation of a score conceived and composed 'without sympathy'.

41, 62
43, 90

'The eye and the ear don't perceive in the same way, and synchronised music doesn't take this into account,' Varèse observes in a 1930 interview on the future development of the movie soundtrack, which he describes as 'the first modern, scientific means music has been given to escape from the tradition in which it is imprisoned.' True synchronisation exploits the human sympathy for events: its ability to adapt itself to Lissitzky's *machine/machine/machine/machine*. 'If the timetable holds good, human laws hold good. The timetable is more significant than the gospels, more than Homer, more than the whole of Kant,' announces one of the characters in Karel Čapek's play, *Rossum's Universal Robots*, first performed in 1921. Metrical simultaneity, as

5, 112
96

Varèse conceives it, organises sound into coordinated sequences. The three-track recording of *Poème électronique* made at the Philips Research Laboratory requires nine separate tape machines operating together in unison, their playback speeds kept as smooth and constant as humanly possible. To be out of synch also means to be out of tune. More significant than the gongs and other exotic percussion on display in his studio is the array of stopwatches photographed on Varèse's desk just after the composer's death in 1965. To lock music into a timetable of specific events requires the most careful synchronisation. Two short compositions are recorded by Xenakis to cover the time taken to reset the equipment between one performance of *Poème électronique* and the next.

· 54 ·

Whenever logical processes of thought are employed – that is, whenever thought for a time runs along an accepted groove – there is an opportunity for the machine.

48, 76   *– Vannevar Bush, 'As We May Think'*

One thing humans generally do better than machines is – adjust. Recorded music can therefore run wild among its listeners in the belief that, concealed by their responses, its effects remain unseen: hence
116   the number of people wearing headphones while on their carefully scheduled way to and from the workplace. The introduction of magnetic tape establishes time as a plastic entity to be measured and cut, spliced together into new sequences and run at different speeds. Half-way between a mechanical device and an electronic medium, the tape
67   recorder works as an intermediate model for explaining the operation of consciousness in time precisely because it has already altered the relationship between them to a profound degree. We exist purely in recorded time, which is also the regime of recorded music, and have no choice but to behave accordingly.

As chance will have it, the American composer John Cage is also at   42, 76
the Brussels World's Fair in 1958 to present a talk entitled, 'Indeterminacy: New Aspects of Form in Instrumental and Electronic Music'. The
lecture is made up of thirty written anecdotes read aloud by Cage
so that each will fit exactly into one minute of elapsed time: the more
words there are in each story, the quicker his delivery becomes. Apart
from this one measure, nothing connects any of the individual anecdotes, either to each other or to the stated theme of the talk. By using
the number of words in each to determine its tempo, however, Cage is
seeking to demonstrate an organisational principle with regard to 'new
aspects of form'. The electronic poem created by Le Corbusier, Xenakis
and Varèse manages to hold together its violent delirium of images and
sounds by exploiting the tension that has always existed between music
and text. Even the 'metrical simultaneity' Varèse speaks of is timetabled
through the simple expedient of being inscribed onto magnetic tape.
For Cage, the text itself is regulated by time to become music. The tension between the two has been subtly shifted, determining the conceptual organisation of the material. It embodies, in an age of recorded
sound, a music whose deepest desire is not to be heard at all.

All art aspires to the condition of music in its wish to endure constant
repetition. Repetition pulls the moment inside out. Repetition makes
meaning total: which is to say, reduced to zero. Repetition is metrical
simultaneity folded back upon itself. 'Record' and 'Playback' constitute
a formula that can be endlessly repeated. Its fullest expression is the
tape-loop. This continuous acetate band elevates sound reproduction into a new form of production: tones, pitches, rhythms and tempi
can all be metrically derived from it. By transforming composition into
a process, however, magnetic tape no longer allows for variations of
interpretation in the manner of a conventional score. Other methods
of notation are required. 'Shall consider further use unfriendly,' Varèse
writes to Cage in 1940 after the latter publicly referred to his own music
as 'organised sound'. The term belongs exclusively to Varèse, the telegram asserts: metrical simultaneity would be unthinkable without it.
Cage's response is to separate simultaneity from synchronisation and
to treat music as organised activity instead.

Once synchronisation is no longer a vital concern, the compositional process becomes a machine for generating individual events or sets of actions. Music transforms itself fully into a model for human behaviour: a temporary Utopia in the making. Created in 1965, Cage's *Rozart Mix* is a work for 'at least 4 performers with at least 12 tape recorders and at least 88 tape-loops', and the score little more than notes exchanged between the composer and Alvin Lucier on the preparation of the loops. The content of each tape-loop is, in every sense of the word, immaterial compared with the independent frequency of its repetition in relation to the others, the state of the machine playing it back and its location in the room. The catalogue describes the piece's duration as 'indeterminate (about 2 hours during first performance)'; and the first live realisation lasts as long as there is an audience to hear it. The free play of multiple tape-loops in *Rozart Mix* restates the relationship between architecture and event as a continuity in which the moment can be extended indefinitely – or at least until the last listener has left the room.

37, 106

• 55 •

To be withheld upon any account, must be a vexation, but to be withheld by a vexation – must certainly be what philosophy justly calls
VEXATION
upon
VEXATION
– Laurence Sterne, Tristram Shandy

'Beethoven was in error,' John Cage declares in a lecture delivered at Black Mountain College during the summer of 1948, 'and his influence, which has been as extensive as it is lamentable, has been deadening to the art of music.' This verdict, the cause of no little friction between Cage and the rest of the college faculty, concerns a fundamental question of structure. 'With Beethoven the parts of a composition were defined by harmony,' Cage argues. 'With Satie and Webern they were defined by means of time lengths.'

Composition conceived purely in terms of duration exposes recorded music as mere data storage; waste and inefficiency are suddenly an issue again. At best, recorded music can only offer consolation for missed experience. The mechanical reproduction of sound also ensures that Beethoven's harmony is transformed into that of bodies and machines working together, unselectively registering sex and death between them. Duration withholds itself from such synchronisation, becoming cumulative in its effects. With duration there are only beginnings and endings – or as Lewis Mumford in his essay 'The Monastery and the Clock' describes it: 'the past that is already dead remains present in the future that has still to be born.' No single event is more important than any other. In a disciplined act of will, Cage manages to replace sex and death with endurance.

29, 63
48, 66

34, 75

It is worth recalling that Cage makes his disparaging comments about Beethoven during a short summer-school celebration of the life and works of Erik Satie. Also present is Buckminster Fuller, who takes time out from his first attempt to construct a geodesic dome with the assistance of some Black Mountain architecture students to play the lead in Satie's music drama *Le piège de Méduse*. The same unadorned simplicity and modulated uniformity that inform the structure of Fuller's geodesic dome can also be found in Satie's later work, in which a hollowed-out aesthetic of monotony, almost mathematical in its regularity, is worked through. Cage later sees Satie's notebooks in Paris and is excited to find clusters of numbers in the margins, which he assumes were used to work out the rhythmical structuring of duration. 'Oh no,' assures the composer Darius Milhaud, who had known Satie personally. 'Those numbers referred to shopping lists.' Even so, the year following the summer celebrations at Black Mountain College, the score for Erik Satie's *Vexations* is discovered and published for the first time. Composed in 1893, the score contains a short musical phrase, barely 80 seconds long, plus the instruction: 'To be played 840 times'. Taking over eighteen hours to perform, *Vexations* is a music that parodies its own wish not to be heard: losing itself instead in familiarity and repetition. In the age of mechanical reproduction, it defies Edison's phonograph to even register its full effect. Unlike Wagner's

42, 78

43, 65

'never-ending melody', whose endless chord progressions appear to lead nowhere, *Vexations* repeatedly returns to its own starting point. In its simple juxtaposition of musical and textual notation, the score represents the emergence not of a new music but a new consciousness.

Like Cage's 'Indeterminacy' lecture, *Vexations* addresses the process by which we adapt ourselves to the inflexibility of the machine. The process itself becomes the machine; and the piano in turn is transformed into an early precursor of the tape-loop. Composition defined in terms of duration invites chance to come into play through the suspension of individual will and preference by creating strategies specifically designed to frustrate them. Duration encourages the withholding of the individual self from the event. In 1963, exactly 70 years after Satie composed *Vexations*, Cage helps to organise its first ever public performance in a Greenwich Village theatre. A small assembly line of pianists is on hand to take over from one another in order that each of the 840 repetitions forms part of a continuous stream. As if to acknowledge that music is no longer a medium through which the relationship between audience and performer is subjected to metrical synchronisation, Cage installs a factory clock in the venue's lobby. Having paid a $5.00 admission fee, members of the public can punch in and out of the event, receiving a nickel refund for every twenty minutes spent listening to the piece. One individual brings a toothbrush and stays for the whole weekend. 'After it was over,' Cage later recalls, 'I drove back to the country and I slept a long time, something like twelve hours. When I got up, the world looked new, absolutely new.'

26, 75

• 56 •

Time didn't have to be measured in meter, but it could be measured in minutes and seconds, and in the case of magnetic tape in inches in space. The common denominator between music and dance is time. This brings up a new situation for dancers.
    – Merce Cunningham, The Dancer and the Dance

Perceived in terms of duration, content is no longer fixed. It becomes a point in time and space: an event. What emerges is something closer to theatre: wildly heterogeneous, interdisciplinary and unpredictable. 'Theatre takes place all the time, wherever one is, and art simply facilitates persuading one of this,' Cage announces. Anything can now be part of a performance. Chance is no longer an abstract mathematical proposition or an expression of malign fate: it has become something far more playful. Cage's greatest achievement may well have been to free the avant-garde from its dependence upon popular culture with regards to form, as evidenced by his collaborations with the dancer and choreographer Merce Cunningham. The simultaneous, the spontaneous and the instantaneous are liberated from the modernist assembly line. Lights, costumes, set design, music and choreography all exist independently of each other, restructuring the event in terms of architecture. There is no single way to experience what is happening.

All action is local – which is to say, vocative. A world that looks absolutely new can only address you directly. At Black Mountain College in 1952 Cage stages 'Theatre Piece No 1', considered the first 'happening' or 'non-dramatic' performance. Cage later writes in the foreword to *Silence* of an audience surrounded by simultaneous activities: 'the paintings of Robert Rauschenberg, the dancing of Merce Cunningham, films, slides, phonograph records, the poetries of Charles Olson and MC Richards, recited from the tops of ladders, and the pianism of David Tudor, together with my *Juilliard* lecture, which ends: "A piece of string, a sunset, each acts."' There are no fixed perspectives on what is happening anymore, no hierarchical ranking of events. With the abandonment of the proscenium arch and traditional sightlines, the mechanical separation of the senses attained in the late nineteenth century is abruptly reversed: 'films, slides, phonograph records' occupy their own space and time. Every seat is the best seat. Each one is supplied with a disposable coffee cup: those who keep their cup until the end of the performance are served coffee afterwards. The programme notes are printed on cigarette papers so that they can be rolled up and smoked. The synchronised harmony of bodies and machines working together is replaced by the individual enjoyment of coffee and smokes.

50, 79

20, 39

30, 68

People are laughing together in a room again. The future is a weightless Utopia of sensory data; 'our work must henceforth consist in brushing information against information', Cage declares in the late 1960s, quoting Marshall McLuhan. With the birth of chance, information is transformed into a sensuous body. Sex and death are replaced by a finite sensuality, reduced to fleeting moments of contact – which is to say, experiences that end. And yet the brushing of information against information is also the beginning of all inquiry. 'The waving of the boughs in a storm,' Emerson observes, 'is new to me and old. It takes me by surprise yet it is not unknown.' At the first performance of John Cage's 4'33", which takes place in August 1952 at the Maverick Concert Hall in Woodstock, New York, the audience finds itself suddenly aware of the wind blowing through the trees outside, brushing information against information, and the patter of raindrops upon the venue's roof. The purest expression of Cage's observation that silence cannot be adequately annotated on a conventional score except as duration, his infamous 'silent' piece is about listening rather than 'not playing' – even though its score comprises carefully-calculated measures and movements indicated by the opening and closing of the piano lid.

• 57 •

Drive car to filling station
Inflate right front tire
Continue to inflate until tire blows out
Change tire
Drive home
(if car is a newer model drive home on blown tire)
        – Robert Watts, 'Casual Event'

The true purpose of revolutionary activity is to establish an environment in which everyone can run wild; stripped of all pretence and representation, humanity returns to a Utopian condition of biological innocence. This, however, can only take place within a technological

regime: one capable of 'the reduction of the traces of sin'. Revolution   34, 107
has nothing to do with civilisation or its discontents – quite the contrary.   15, 45
The return of the repressed is, and can only be, the total imposition of
the repressed. 'The hidden aim of technological progress is the discovery and recovery of the human body,' Norman O Brown observes in *Life*   40, 74
against Death, a book mentioned repeatedly by Cage in his writings.
Technology, in other words, is the institutionalisation of the wilderness:
an extension of the sensuous body. Information brushes against information, in Cage's words, 'like a suit of clothes'. Part of the very fabric of
an absolutely new world, information behaves like an element, defined
by the structures it generates from moment to moment as it comes into
fleeting contact with itself.

Cunningham's choreography, according to Cage, deals directly with
the problem of 'assembling heterogeneous facts that can remain without interrelationships'. The simplest way of grasping 'the possibilities of
the random situation' is to walk through the city, a detached observer
of what is happening around you. 'The city as theatre,' Cage observes,
'helps show people this potentiality.' As a corporate experience, the city
recombines our mechanically-separated senses in radical new ways – it
has no choice. Built upon the surplus value generated by the agricultur-
al exploitation of the wilderness, the city is a sacred space devoted to
the worship of information as an environmental element – or, to express
it in more intimate terms, our 'second nature'. A former student of architecture, the young Cage dismisses Le Corbusier's call for modular
proportion in favour of 'the clutter of the unkempt forest'. The sound of
fridges and humidifiers, passing traffic and voices becomes 'like sculp-   109
ture or like a definition of space' to him. Architecture is no longer simply
a frozen music; the city is made 'audible to itself', taking on the form and
aspect of a graphic score composed in three dimensions. 'The more
glass, I say, the better,' Cage declares. The open planes, glazed spans
and steel frameworks of modern building design no longer separate
public space from private, up from down, inside from outside, city from
wilderness. 'A piece of string, a sunset, each acts.' All is information acting with information now – no more, no less.

26, 65   Cage's graphic scores are events in the making: conceptual machines created from codes and instructions, tables and transparencies, grids and overlays designed to externalise thinking by placing it in the hands of others. Performers realise each new composition by making themselves a part of the process it initiates. As information becomes layered upon information, chance comes into play. 'What I am calling poetry,' Cage remarks, 'is often called content. I myself have called it form. It is the continuity of a piece of music. Continuity today, when it is necessary, is a demonstration of disinterestedness, that is, it is proof that our delight lies in not possessing anything.' That which can no longer be regarded as a possession exists solely as an event. Meaning becomes a

24, 93   technological environment established by those who participate in it: one capable of producing unexpected thoughts and actions.

• 58 •

Ptr .. r .. rr.. ing – twing-twang-prut-trut – 'tis a cursed bad fiddle – Do you know whether my fiddle's in tune or no? – trut.. prut.. – They should be fifths. – 'Tis wickedly strung – tr… a.e.i.o.u.-twang – The bridge is a mile too high, and the sound-post absolutely down, – else–trut .. prut – hark! 'tis not so bad a tone. – Diddle diddle, diddle diddle, diddle diddle, dum. There is nothing in playing before good judges…
    – Laurence Sterne, Tristram Shandy

'Ideas are one thing,' Cage observes, 'and what happens is another.' Chance intervenes. The metrical simultaneity of the 'electronic poem' created by Varèse and Le Corbusier in 1958 has been rewritten as free verse, utilising the syntax of equal emphasis – which is to say, equal significance. Sterne's transcription into words of a violin tuning up ruins the syntactical flow of the text: a sequence of vowels and random noise replace the actual tuning of the instrument. Words, noises and silences, which are the pauses between utterances, are all that remain. *Poème électronique* is already a museum piece before the Philips Pavilion even opens its doors. What has taken its place is a form of play: the parody of civilised behaviour. For the world premiere of his *Concert for Piano and Orchestra,* which also takes place in 1958, Cage transforms Merce Cunningham into a human chronometer. 'I had to wear a rented dress suit, tails and everything,' Cunningham later recalls. 'John wanted the conductor's arms to represent the hands of a clock. I think we only did it once, but his idea was that at each performance the tempi would change. The musicians all had a score, which contained their individual parts. OK. If you were looking at the conductor...' Cunningham's left hand goes up. 'It was all done with your arm going clockwise.' His arm deftly marks out the quarter-hours. 'This is fifteen, this is thirty, this is forty-five, this is sixty. Say that this fifteen minutes is to go slowly, taking a half-hour, the conductor would be moving his arm slowly like this. The players realise it's slow and change their tempo. Of course John went way beyond that. He usually didn't want a conductor at all.'

In 1961 George Maciunas, head of the Fluxus group in New York, relocates to Germany to work as a civilian employee for the US Air Force. In 1962, Nam June Paik, who has been working with Karlheinz Stockhausen in the electronic music studios at WDR Cologne, stages 'Solo for Violin' in which a violin is slowly smashed to pieces. In 1963 a two-day 'Festum Fluxorum Fluxus' dedicated to the spirits of 'music and anti-music' is held at the Düsseldorf Kunstakademie; participants include George Maciunas, Nam June Paik, Dick Higgins, Robert Watts, Jackson Mac Low, Terry Riley, György Ligeti, La Monte Young, Henry Flynt, John Cage, Yoko Ono, Joseph Byrd, Cornelius Cardew and Brion Gysin. Also taking part but not listed on the festival poster is Joseph Beuys, the academy's newly appointed professor of sculpture, who performs *Siberian Symphony, First Movement* with a blackboard, a dead hare and music by Erik Satie played on a piano prepared with lumps of clay. A second piece, lasting only twenty seconds, is *Concert for Two Musicians:* a windup toy of two clowns crashing cymbals to-gether is allowed to run down in front of the audience.

39, 75

94

4, 63

A mixture of scripted confrontations, spontaneous provocations and staged interventions, Fluxus events are partially inspired by Cage's idea of music-making as a form of Utopian social activity. The distinctions between form and content, performer and audience, action and representation become wilfully confused at events in which conventional instruments are systematically abused, household objects and electronic devices used to generate sounds, and the duration of individual pieces is dictated by chance or circumstance. 'The aims of Fluxus are social (not aesthetic),' Maciunas writes in 1964, 'and are directed to step by step elimination of the Fine Arts (Music, Theatre, Poetry, Prose, Painting, Sculpture etc). This motivates the desire to redirect the use of materials and human ability into socially constructive purposes: in other words the applied arts, industrial design, journalism, architecture, science, graphics and typography, printing etc... all of which are related to the fine arts and offer fine artists the best alternative working possibilities.' If civilised behaviour is something to be overcome, might not music have to suffer the same fate? It is not life without music but music itself that is the mistake, the correction of which can rarely be more than hypothetical or temporary at best. Does anyone still believe what they are hearing? 'The servitude of man to man and the enslavement of man to matter will cease,' Čapek predicts in *Rossum's Universal Robots.* 'Of course, terrible things may happen at first...'

20, 63

· 59 ·

Revolutionary culture is a powerful revolutionary weapon for the broad masses of the people. It prepares the ground ideologically before the revolution comes and is an important, indeed essential, fighting front in the general revolutionary front during the revolution.
  – Mao Tsetung

'Current psychoanalysis has no Utopia,' Brown declares in *Life against Death,* 'current neo-orthodox Protestantism has no eschatology.' Both, being modern, express no greater aim than to prevent 'terrible things' from happening. As a youth, Cage's ambition is to become a Methodist minister; as an adult, he visits a Jungian analyst in an effort to save his marriage. 'I'll be able to fix you so that you'll write much more music

than you do now,' the analyst assures him. 'Good heavens!' Cage replies. 'I already write too much, it seems to me.' The composer who
smokes three packs a day, can't imagine a meal without butter or
cream, drinks wine by the bottle and has sex with men disappears into
the physical directness of his music: 'each acts'. The unseen harmony,
Heraclitus reminds us, is stronger than the seen. In the early 1970s Cage
starts quoting Chairman Mao alongside Buckminster Fuller, looking for
connections between the two.

The prospects before humanity, as presented by Fuller in 1969, remain   42, 78
clear: *Utopia or Oblivion*. In discovering and recovering the body, the
true hidden aim of all technology is to become indistinguishable from
nature. Technological advance is often perceived in cultural terms as
the reintroduction of the primitive into society. Just as culture seems
closest to nature when it appears to impede progress, so the natural
must assume the features of an intense spirituality in order to reassert
itself. Unlike spiritual experience, however, technology does not need
to be relearned personally. Expressed in Utopian terms, it transforms
the sensual life of the human animal into a form of social architecture.
The sound-houses of *The New Atlantis* stand open before us.

The pleasures aroused by sound, Marcuse notes in *Eros and*   28, 74
Civilization, seem 'more sublime' when compared to the 'bodily' ones
associated with touch, taste and smell. Hegel defined sound at the
dawn of the nineteenth century, as 'a disappearing of being in the act
of being'; by century's end, the phonograph suspends that moment
of disappearance forever. 'In the phonograph we find an illustration
of the truth that human speech is governed by the laws of number,
harmony and rhythm,' Edison writes in 'The Perfected Phonograph'.
The mass revolution that makes speech 'immortal' also enforces the
unseen harmony of mechanically reproduced sound. 'Style, rhythm,
meter introduce an aesthetic order which is pleasurable: it reconciles
with the content,' Marcuse observes in a chapter appropriately titled
'Phantasy and Utopia'. With the advent of recorded sound a tension
develops between the unseen harmony of music, whose invisible
dimension is revealed to be technological in nature, and its deepest

desire not to be heard at all, which must finally do away with all pretence of representation. The birth of chance comes from the spirit of music – at the end of music. 'Art survives only where it cancels itself, where it saves its substance by denying its traditional form,' Marcuse continues – otherwise, 'it shares the fate of all genuine human communication: it dies off.'

· 60 ·

I tell you before-hand, you had better throw down the book at once; for without much reading, by which your reverence knows, I mean much knowledge, you will no more be able to penetrate the moral of the next marbled page (motly emblem of my work!) than the world with all its sagacity has been able to unravel the many opinions, transactions and truths, which still lie mystically hid under the dark veil of the black one.
    – Laurence Sterne, Tristram Shandy

Acceptance weakens an idea. If nothing else, Fluxus offers a brief moment of escape from music's invisible dimension: a silencing of its unseen harmony. In one playful gesture the perceivable is wrenched from the intangible; but the moment is fleeting. What remains are the announcements, posters and documentation: events, actions and happenings ultimately gain permanence as works of graphic design. The conceptual artist becomes a creative writer: Sol LeWitt, Joseph Kosuth, Robert Smithson, Joseph Beuys and Nam June Paik start to create their own forms of fiction: ad hoc assemblages of propositions that constitute individual universes composed of signs and spaces. Just as Sterne inserted pauses and omissions, black, blank and even marble pages into Tristram Shandy's rambling narrative, so in his lectures Cage 'left several pages and parts of pages blank – visual conversions of long musical rests meant to be treated as pauses when reading the lecture aloud'.

Essays and lectures about music and indeterminacy are gradually replaced by texts that convey nothing beyond their typographical presence on the page: they become the embodiment of an idea instead of

its expression. 'Typographic changes, like the "mosaic" form, are noises which erupt in the book!' Cage observes of this process. 'At one and
the same time, the book is condemned to non-existence and the book
comes into being. It can welcome anything.' To drive meaning out of
the text is, in his view, also to discover more music in the world, 'once all
psychoanalysis has been pushed aside'. In the guise of Tristram Shandy
'speaking of my book as a *machine*', Sterne connects consciousness
with the erotic distractions, conversational digressions and syntactic   44, 64
inefficiencies that Freud will later refuse to allow his patients to inscribe   15, 46
onto the phonograph. The 'typographic consciousness' which permit-
ted Sterne to perceive text as an unstable machine made up of words
is barely even acknowledged in a culture now attuned to mechanically
reproduced sound. 'As soon as you surpass the level of the word,' Cage
is able to declare, 'everything changes: my essays...didn't deal with
the question of the impossibility or possibility of meaning. They took for
granted that meaning exists.'

Music is the corruption of language, and Wagner is as much an   48, 65
inevitable product of this process as Cage. Poetry began as a social utility, a communications medium for the supplication of the gods; music
would find its own origins in those same rhythms and repeated sonor-
ities. At the same time the typographical arrangement of the alphabet
completed what McLuhan describes as 'the translation or reduction of
a complex organic interplay of spaces into a single space'. In linking the
body to this single space, text becomes a machine. Language reconfig-
ures itself, in Norman O Brown's words, as 'an operational structure on
an erotic base'. Its seemingly inevitable corruption into music offers a
brief refuge from meaning; a chance for the body to return to a Utopian
condition which the graphic score helps to restore in a new form. 'With
such a transfigured body,' Brown reveals, 'the human soul can be reconciled, and the human ego become once more what it was designed to
be in the first place, a body ego and the surface of a body, sensing that
communication between body and body which is life.'

· 61 ·

The rationalization you call for has a great deal of superstition about it –
of a belief in something impalpable and vaguely demonic that's more
at home in games of chance, in laying cards and casting lots, in augur-
ing. Contrary to what you say, your system looks to me as if it's more
apt to resolve human reason into magic.
        – Thomas Mann, Doctor Faustus

Beethoven took the general principles of composition, those same
laws of number, harmony and rhythm Edison later identified as governing human speech, and placed them in the service of individual
expression. A piece of music was no longer the restatement of a
universal aesthetic order already determined by key, meter, phrasing
and form but would come directly 'from the heart' instead. Beethov-
en also preferred to describe himself as a *tondichter* – or 'sound poet'
– rather than a composer, suggesting that all art aspires primarily not
to the condition of music but to that of language – which makes more
sense. Cage returned composition not to the universal laws of harmony
and whole numbers but to the fractional arithmetic of chance: to the
randomised methods of operation to be found in nature. The psychology of individual expression is replaced by the exploration of space and
process: which is to say, by the uniqueness of the event itself as defined
by its duration.

Cage's pronouncements on language remain sparse, simple and at
times painfully naïve, but they are also persistent, demanding and absolute in their claims upon our attention: 'Utopian' even to the point of
arguing for a society that has no need to communicate at all. Language
often appears as a physical entity; he speaks of 'raising language's
temperature' or of it becoming 'arthritic' as it conforms to the laws of
syntactic organisation. 'The polymorphous perverse: the necessity of
Utopia,' he notes, referring to the phonetic density of puns, the typographic complexity of overlays, random superimpositions, elisions that
free language from the rule of syntax, promoting silence and nonsense

in its place. Text becomes, for him, the final expression of music's secret desire not to be heard; it represents a music uncorrupted by its compulsion to seduce through illusion while remaining unseen. Perhaps, in the end, Cage respects text too much; his 'writing through' *Finnegans Wake*, a series of deconstructions based upon the spelling out of James Joyce's name, adds little to an understanding of the novel – except perhaps as an oblique reminder of why the author chose to call his first collection of poems *Chamber Music*. In later life Cage becomes 'stuck in the *Wake*', quickly finding he that 'couldn't get out'.

The text becomes a wilderness; and the typographical reorganisation of bodies in space is transformed into choreography. 'The Cretan labyrinth was Ariadne's dancing ground,' Norman O Brown reminds us   5, 49
in *Love's Body*. 'After their victory, Theseus and the young Athenians danced a dance consisting of certain measured turnings and re-turnings imitative of the windings and twisting of the labyrinth.' The time has come to apply their rhythms to thought itself. Music is as permanent a part of our environment as the human form and is as equally mortal. The more it combines itself with interactive material, contextual statements, packaging and audiovisual data, the more the possibility arises of music existing only as a set of organisational principles. Seen from this perspective, it may be that the greatest work by the German composer Adrian Leverkuhn in Thomas Mann's *Doctor Faustus* is the one he will never hear: his own life 'as told by a friend', which is the text of the novel itself. In structural and thematic terms, *Doctor Faustus* and *The Life and Opinions of Tristram Shandy, Gentleman* are both the work of unreliable narrators; themes are introduced far too early and resolutions often come too late. The words flash before the reader's eyes: the interior and exterior spaces of the Labyrinth collapse into one   40, 72
another as each narrative inevitably folds back on itself. Utopia remains a narrative fragment, an ephemeral collection of sights and sounds and gestures that takes place in the distended present of text. Beyond that it can only reside in some hypothetical future where people will always be laughing together in a room somewhere.

Invading Present Time:
The Politics of Simulation

'The *task of assimilating knowledge* and making it instinctive
is still quite new; it is only beginning to dawn on the human eye
and is yet barely discernible – it is a task seen only by those who
have understood that so far we have incorporated only
our *errors* and that all our consciousness refers to errors!'

Friedrich Nietzsche
The Gay Science

· 62 ·

No, 'They' are not God or super technicians from outer space. Just technicians operating with well-known equipment and using techniques that can be duplicated by anybody else who can buy and operate this equipment.
– William S Burroughs, 'Electronic Revolution'

In 2008 a BBC radio announcer offers a personal apology to the family of deceased Hollywood screenwriter Abby Mann for having dissolved into uncontrollable laughter while relaying details of his death on the morning news. This inappropriate outburst had been provoked by the item immediately before it – a recording of a human voice made in 1860 had just been successfully reconstructed in a California laboratory, making it the earliest in human history. Created by French typesetter Édouard-Léon Scott de Martinville on a device of his own invention called the 'phonautograph', it had existed for almost 150 years as little more than the representation of a sound wave etched into a piece of soot-covered paper. Played back digitally in the twenty-first century using a 'virtual stylus', what had originally been a graphic inscription of the folk song 'Claire de Lune' now sounded in a BBC radio studio more like 'a bee buzzing in a bottle', sending the newsreader into a barely-suppressed fit of the giggles. You can appreciate why. Everything about the incident works towards provoking precisely this response: a spectral set of associations holds the moment together. There is the barely discernible

45, 91

voice retrieved from a smoky deposit with its accumulated cloud of a song; and the dead Hollywood writer's obituary with its sense of mortality and time passed. More importantly a historical overlay of different media has briefly been established, from graphic inscription to script,

53, 86

from song to mechanical recording, from text to cinema and broadcast radio. 'The fact is it's recorded in smoke,' exclaims the sound historian who had discovered the original 1860 recording preserved on a strip of rag paper in the archives of the French Academy of Sciences. 'The voice is coming out from behind this screen of aural smoke.' Except that 'Claire de Lune' has not been played back, but a form of 'playback' has been digitally simulated instead.

99
53, 77

Just as the written instructions that make up the Turing Universal Machine allow it to simulate the behaviour of every other data-processing machine, including itself, at the heart of each computer, so the 'virtual

30, 87

stylus' simulates playback of a voice recorded in smoke. Like a ghost, it becomes an event: intrusive but without limit or dimension. To laugh, under such circumstances, is a way of expressing incomprehension, which also covers for a reluctance to believe. We see the same smoke emerge a century later at the beginning of *The Testament of Orpheus*, Jean Cocteau's last film, released in 1960. '*A tangle of smoke dissolves*

42, 90

slowly and curls out through a soap bubble, that appears to have come out of the point of a knife,' the screenplay stipulates, as a new form of audiovisual poetry is brought mechanically to life. This opening effect is created by the simple expedient of running the film backwards: an event that betrays its own origin. '*My hand, the knife, the bubble leave the screen,*' Cocteau's script then reveals, '*and give way to an empty studio where all the elements of the set will appear in succession.*'

Right from the beginning we have mistrusted all representations of the world; and when we seek to live through representation alone, we come to mistrust them the most. In the writing of *Tristram Shandy*, Sterne produced an engine that generated itself – or rather, it generated a hypothetical relationship between the book's supposed author and his creator based on simulation. The concluding film in Cocteau's 'Orphic Trilogy', *The Testament of Orpheus* is presented by its creator as 'simply a machine for creating meanings'. Simulation becomes the event itself: the film is run backwards and forwards, sudden edits make characters appear and disappear. Thanks to such audiovisual tricks, Cocteau is no longer a poet but a ghost trapped inside a film studio, moving in and out of present time, in search of his own death.

· 63 ·

Analyzed correctly, our revolution is a revolution of grammarians who battle and cut each other's throats in order to dethrone words, and our liberty resembles to a T those magicians in fairy stories who would overthrow nature by pronouncing a few quaint expressions.
   – La Quotienne, 1797

While checking the ventilator in his restaurant, The Thousand and One Nights of Tangiers, one afternoon in late 1957, Brion Gysin finds a charm   4, 58
made up of mirror shards and seeds, stuck together with various bits of biological goo. Inside it is a text laid out in a cabalistic grid. 'May Massa Brahmin leave this house as the smoke leaves this fire, never to return,' the invocation read. A few days later 'Massa Brahmin' is, by his own account, out of the restaurant business, out of Tangiers 'with the shirt on my back' and heading for Paris and lodgings in a cheap hotel. It is here, in the cold spring of 1958, that he chances upon the 'cut-up technique' while slicing through a pile of newspapers with a Stanley blade. Along the axis of each cut, the layers of printed text form themselves into sequences of randomly juxtaposed words from which new meanings emerge. Busily lashing together the manuscript of *The Naked Lunch* in the room below, William Burroughs hails Gysin's discovery of the   20, 84
cut-up as initiating a 'project for disastrous success.'

The rest of the story writes itself. 'On the magical plane,' according to Adorno and Horkheimer, 'dream and image were not mere signs for   40, 70
the thing in question, but are bound with it by similarity or by name.' The relatedness of things, from which the earliest forms of poetry are derived, eventually gives way to the politics of intention, however.

'Like science, magic pursues aims,' Adorno and Horkheimer conclude, 'but seeks to achieve them by mimesis – not by progressively distancing itself from the object.' Magic, in other words, is sexualised action. Poets have always understood this better than most. Tristan Tzara observed that 'thought is made in the mouth', while it was human breath that determined the length of each line in Charles Olson's verse. Both ultimately saw poetry as a typographical act – for Olson, the technology of the typewriter, due to the rigidity and precise spacing of each letter, allowed him to indicate pauses, the suspension of syllables and the juxtapositions between parts of phrases. Poetry is created not with ideas but with words, Stéphane Mallarmé affirmed. To go further and see poetry made up from twenty-six letters is to glimpse the remains of the magical formulae that were once hidden within it.

'There is no essential difference between a human brain and a machine,' declares a Futurist manifesto hammered out by Bruno Corradini and Emilio Settimelli in 1914. 'It is mechanically more complicated, that is all. For example, a typewriter is a primitive organism governed by a logic imposed on it by its construction.' Having established that all human activity 'is a projection of nervous energy', the manifesto equates the basic physiological components of poetry with those of the typing machine. 'A broken key,' claim Corradini and Settimelli, 'is an attack of violent insanity.' Typographical text constitutes a biomechanical entity: one that is constantly poised to invade present time. Both a tool and an instrument, the typewriter transforms the arrangement of words into a grid to be manipulated. What once formed the basis for magical practice now offers itself up to being sliced through, collaged and juxtaposed. Texts are therefore edited into existence, both as part of the world experienced through the senses and as something quite apart from it. Typesetting, together with the possibility of juxtaposing text with text, or text with image, marks the end of handwriting and the beginning of graphic design. The cut-up technique, as applied by William Burroughs to his own writings, becomes a more violent expression of the editing process: a breakthrough that looks forward to the point at which text, sound and image are no longer separated from each other but converge in the pale glow of a laptop screen. The issue

55, 79

58, 75

52, 85

is no longer how something functions but what effect it has. Remington, the company that produces the UNIVAC mainframe computer and manufactures the first typewriter to feature the QWERTY keyboard, 38, 77 starts out making guns and ammunition. It is only when the demand for weaponry recedes at the end of the American Civil War that Remington moves into the business of typewriting.

· 64 ·

For science the word is a sign: as sound, image and word proper it is distributed among the different arts and is not permitted to reconsti- 22, 78 *tute itself by their addition, by synesthesia, or in the composition of Gesamtkunstwerk.*
        – Adorno and Horkheimer, Dialectic of Enlightenment

The 'original sin of art', as laid down by Cocteau in his preface to *The Testament of Orpheus,* is that it 'wanted to convince and to please'. Magic represents the eroticisation of spiritual agency precisely 60, 79 because it ignores the very restrictions and boundaries that have made the joint acts of convincing and pleasing so necessary in the first place. Treated as modern forms of magic, the electronic media of the 1960s invade present time, distracting everyone by locking them into the immediate. *'consider this machine,'* Burroughs writes of the tape recorder in his essay 'the invisible generation', typed out entirely in lower-case letters, *'it can record and play back activating a past time set'* – like smoke, poetry becomes an amorphous audiovisual presence drifting in from another dimension: *'what we see,'* Burroughs stipulates, *'is determined to a large extent by what we hear'* His experiments with typewriters now involve tape recorders as well. Extending the application of Gysin's cut-up technique to include both typographical and recorded words, the novels he produces immediately after *The Naked Lunch* throw meaning back upon itself. *The Soft Machine, The Ticket That Exploded* and *Nova Express* go to great lengths to scramble the logical organisation of words and their relationship with the world. Upsetting the established concepts of what is expected from fiction in order for it to please and convince, this delirious sequence of fictions is a perfect expression of the prevailing Electronic Delirium, aggressively exposing how time, memory and consciousness no longer form a smooth continuity – especially in a culture mediated by the tape recorder and the television camera. 'As the preliterate confronts the literate in the postliterate arena,' remarks McLuhan in 1969, 'as new information patterns inundate and uproot the old, mental breakdowns of varying degrees – including the collective nervous breakdowns of whole societies unable to resolve their crises of identity – will become very common.'

51, 80
60, 96

Through the mediumship of the tape recorder, Burroughs intersects 'literate' typographic text with radio and TV voices, the angry noises of the street and the grotesque banalities of modern consumer culture. Massed presences swarm through his work: *the invisible brothers are invading present time,'* he writes in 'the invisible generation'. Invading present time was never going to be either a systematic or an orderly affair; as a process it has all the turbulence, the formless violence, of smoke. The manipulation of word and image, juxtaposing recordings of past and present time can, Burroughs argues, profoundly alter consciousness, influencing and controlling perceptions. 'Lee awoke with his spine vibrating and the smell of other cigarette smoke in his room – He walked streets swept by colour storms slow motion in spinal fluid came to the fish city of marble streets and copper domes – Along canals of terminal sewage – the green boy-girls tend gardens of pink flesh – Amphibious vampire creatures who breathe other flesh…' All juxtaposition is violent – it can be nothing less. 'What precisely is a dream?' Burroughs muses at the start of a 1967 interview before concluding: 'A certain juxtaposition of word and image.'

53, 84
13, 93

· 65 ·

What we are tracking here is: How does a word become a picture? and, How does a picture become a word? In either case, you know it is happening when something clicks. For a picture, the click is like a camera shutter. For a text, it is more like the click of a tape-recorder switch. Listen for that click.
    – William S Burroughs, 'Precise Intersection Points'

The liberation of the voice as incantation through the inscription of words onto magnetic tape is just one more in a series of radical tilts, reversals or inversions that have occurred in human communication since handwriting first emerged in Mesopotamia over 5,000 years ago. It was in this region that cylinder seals were first used from 3300 BC onwards to create a raised relief on soft clay as an identifying mark. Deities, animals and scenes from daily life rose up from the dampened river mud: tablets dating as far back as 3100 BC record the accumulation of business transactions over long periods of time, involving different cities from Iraq to Iran. In a further innovation, these transactions are recorded as pictograms representing animals and objects inscribed into the clay with a stylus instead of being pressed out of it under the rolling motion of a cylindrical seal.

25, 80

Bound by similarity, the pictogram remains an invocation of the thing it represents. Simplified as script, however, it loses all pictorial power: in other words, its content. The transition from pictogram to a sign capable of representing sounds or abstract ideas is marked by the former being tilted to one side. On the flat plane of a wax tablet, the original pictogram undergoes a ninety-degree declination, indicating that it should now be viewed differently; no longer something to be looked at, the sign must be read instead – mimesis has been replaced by meaning. Later standardisation established a written language that has lasted for over 3,000 years – but it is the inversion involved that has brought this about. The first impression made in softened wax by a stylus inverts the action of the cylinder seal, which produces figures in relief out of damp clay; another closely related inversion takes place when Edison uses a

60, 75

55, 96

stylus to inscribe the sound of a human voice into the softened wax of a recording cylinder. Speech, once thought incapable of restraint, now has the permanence of writing.

Most ancient civilisations leave their words and pictorial forms behind them but rarely their actual sounds. The self-conscious revaluation of culture starts with the recording of speech and music: from paper to shellac to acetate to plastic to silicone, our world has grown increasingly heavier with it – even as its presence, transformed into digital code, seems lighter and more accessible than ever. In the meantime human history continues to be recorded by obsolete devices and outdated technology.

57, 75

• 66 •

Speech, though it deludes physical force, is incapable of restraint.
– Adorno and Horkheimer, Dialectic of Enlightenment

It is a small step from pleasing and convincing to deluding: no wonder we habitually mistrust all representation. You may only see that which is directly in front of you; but you hear what's coming from a distance. The phonographic reproduction of speech abruptly separates hearing and sight, creating further reversals and inversions. Together or apart, the two senses constitute an audiovisual fault line from which more fractured simulations of experience emerge. It is through the basic discrepancy between perceptual fields – between hearing and sight – that ghosts start to creep in. Consequently, no medium proves itself until it has been effectively used to communicate with the dead. Spirits rap on darkened tables in Morse code; Marconi and Oliver Lodge claim that their discoveries can help the living contact the dead. Death, like hearing and sight, becomes a sensory affair, involving a simple change in the human wavelength. Recorded sound is thus the shed skin of a world that had previously been experienced through the full bodily interplay of the senses; as such it brings about a hardening and hollowing out of experience.

79, 91

In becoming immortal, recorded speech is rendered flat and two-dimensional but possessing a presence that had never existed before. The machine itself appears to be speaking your own words back to you. Immortalised speech consequently finds itself subject to restraint: its permanence takes the form of inflexible repetition. In an effort to perfect this illusion Edison develops his improved Blue Amberol cylinder machine while at the same time announcing plans to work on a device that can be used to communicate with the dead. Through the standard-isation of production and parts, an assembly line of more complex and delicate mechanisms becomes possible; the introduction of increasingly precise machines inevitably leads to increasingly precise repetitions. Simulation requires the smooth running of rhythmic engines, driving out all imperfections and inefficiencies in the form of ghosts that can no longer be mistaken for anything else. Every advance in the technology of sound reproduction shows up the inefficiencies of what went before, isolating them as phenomena in their own right. What sounded to earlier ears like a perfect and faithful playback now seems alive with scratches, hums and clicks. The signal-to-noise ratio shrinks perceptibly; and the newly restricted audio range is soon buzzing with unexpected messages. Electronic Voice Phenomena, or EVP, is the term used to cover a wide range of sounds, utterances and intrusions of unknown origin that have somehow managed to register themselves upon our consciousness through existing communications channels.

53, 98

Many consider EVP a means by which the dead speak to the living. It is only as recordings that they come into existence. Immortalised speech, still subject to the laws of repetition, has shaken off the restraints imposed upon it. The dead display a meaningful lack of consequence, however. Their words seldom make any sense, often taking the form of riddles, veiled allusions or cryptic asides. Try listening to Edison recite 'Mary had a little lamb', still the first recorded words ever to be played back successfully on a mechanical device: the shock of his voice reaches across the years. That humans and machines might interact with each other to produce some inhuman communication remains an intriguing proposition. For writers like Cocteau and Burroughs, it opened up the possibility that a new kind of audiovisual poetry might

119

emerge – or at the very least be better understood. Meanwhile Edison's choice of text remains as cryptic and perplexing as any spirit message. Caught somewhere between a nursery rhyme, a song and an engineering test, it invokes a phantom presence that has been with us ever since. Speech has escaped its boundaries once more. 'And everywhere that Mary went... the lamb was sure to go.'

29, 98
44, 114

· 67 ·

in a modern house every room is bugged recorders record and play back from hidden mikes and loudspeakers phantom voices mutter through corridors and rooms word visible as a haze
  – William S Burroughs, 'the invisible generation'

Burroughs may well understand how sordid the concept of progress can be; yet he still believes in it. He is, after all, writing in, and for, an age that regards having a tape recorder in the home as an act of self-improvement. Unlike radio or television, it is one of the few pieces of electronic equipment around the house with which you could interact fully; and unlike the telephone and the hi-fi system, the tape recorder puts you completely in charge of what you are hearing. From the 1950s onwards, there is a tendency to treat magnetic tape as an analogue for human consciousness as it exists in time. Individual experience becomes a recording that can be rewound, reviewed and, if necessary, erased and rerecorded. Behaviourists, psychotherapists, LSD researchers and even Scientologists – with whom Burroughs enjoys a brief dalliance – begin to examine how the human time track can be played back, spliced and edited.

54, 102

Describing the tape recorder as an extension of the central nervous system, Burroughs's recommendation that his readers tape their experiences in order to *'get it out of your head and into the machines'* echoes Edison's remark that 'the phonograph knows more about us than we know ourselves'. This transfer of awareness reveals how little we actually perceive; to play back events recorded on tape, Burroughs argues, is to discover that *'there was a grey veil between what you saw or more often did not see that grey veil was the prerecorded words of a control machine'.* Consciousness itself is manipulated through the senses, which in turn become reconfigured as extensions of the Electronic Delirium. 'So Who Owns Death TV?' Burroughs asks from the cover of a small press pamphlet. We all do, it seems. A believer in progress, he may call the pre-recorded universe into question, but he never doubts the reality of the playback equipment itself.

Recorded sound links existence with nonexistence, the animate with the inanimate; other recordings may already exist which we are not as yet technologically equipped to play back. 'Remember that when the human nervous system unscrambles a scrambled message,' Burroughs observes in 'Electronic Revolution', 'this will seem to the subject like his very own idea which just occurred to him, which indeed it did.' Thanks to the surface noise of contemporary mass media, a world of unseen events has become an integral part of our daily reality. Privately printed in 1971, 'Electronic Revolution' is a tactical treatise on how to use sound and image to trigger profound cultural upheavals. That same year Dr Konstantin Raudive also publishes *Breakthrough* on his researches into EVP. Its accompanying vinyl disc contains recordings of unknown voices from a hidden dimension, speaking in polyglot tongues, expressing themselves in the same oblique, fractured poetry that Burroughs obtained from Gysin's cut-up technique. Whatever their origin, the tone of these messages is often sarcastic, elusive, ominous, even threatening. Voices from the dead invading present time, they offer cryptic warnings on the sorry state of human affairs, hinting darkly at future events. Furthermore they express no doubt whatsoever over who is in charge around here. 'We work the switches,' a mysterious radio voice informs an EVP researcher in no uncertain terms the year after *Breakthrough* first appears.

· 68 ·

Kant foretold what Hollywood consciously put into practice in the very process of production: images are pre-censored according to the norm of understanding which will later govern their apprehension.
  – *Adorno and Horkheimer,* Dialectic of Enlightenment

95

George Romero's *Dawn of the Dead* opens on scenes of chaos inside a local television station. Technical staff and presenters run around in panic, abandon their posts, throw their scripts and production notes angrily into the air, letting the pages fall where they may. Cocteau's empty studio is now filled with the voices of angry ridicule and despair. Zombies are taking control of the city while civil order collapses. To keep people watching even in these final days, the channel continues to broadcast the addresses of inoperative rescue stations, effectively sending its viewers to their deaths. 'Our responsibility is over,' a camera operator dreamily remarks before going back to work. Reality is a film over which we have little direct control. 'The need which might resist central control,' Adorno and Horkheimer argue, 'has already been suppressed by the control of individual consciousness.' Burroughs conjures up 'the

14, 78

reality studio' at the height of the Electronic Delirium to illustrate the proposition that all human experience has been programmed before the fact: most of what we see or do is pre-recorded. By the time we realise that all the rescue stations are inoperative it may already be too late.

Hollywood revels in chaos: the end of the world presented in, and as, an empty movie studio – 'no disaster film,' declares the German director Hans-Jürgen Syberberg, 'but disaster as a film'. We apprehend the universe by extending our original five senses: our sixth sense is boredom.

30, 86

The true imagination of disaster is the one that secretly wills such events to happen; whether the product of electromagnetic pollution or subconscious apprehension, EVP is the advance signal for a society ready to self-destruct. The energy of mechanical sound is not constant but, like sexual energy, renews itself through rhythmic repetition. Running a recording backwards and forwards, randomly cutting in new words and sounds, alters its predetermined dynamics, allowing unfamiliar

events to reconstruct themselves from the electromagnetic silence. 56, 81
The voices of the dead establish an anti-environment in which we
become alienated from the pre-recorded world around us. The newsreader's laughter fills the airwaves. 'However, it's a lot easier to start
trouble than to stop it,' Burroughs dryly remarks as we finally prepare to
take leave of our senses.

Suddenly ghosts are no longer enough; it's the total possession of the
medium itself that is important – a rendering of human consciousness.
'How random is random?' Burroughs asks, referring explicitly to Raudive's tape experiments. 'We know so much that we don't consciously
know, that perhaps the cut-in was not random.' At the end of an interview in 1970 he observes that 'if everything is an illusion, then everything
is permitted. When things become real, definite, then they are not
permitted. Now our culture, by and large, absolutely reverses that –
everything is true and nothing is permitted. That is the whole stand
of the reactionary Establishment: make everything true and permit
nothing.' Illusion, Burroughs concludes, is a revolutionary weapon that
can create unrest, provoke riots, spread false rumours, discredit news
organisations and politicians. Control is revealed as a pre-recorded
mirage: an encoded social impulse played back on streets and movie
screens, in cities and living rooms, across newspaper headlines and television channels. The scrambling together of mind and machine brings
about the end of the human psyche as a cohesive whole, leading in turn
to the diffusion and disappearance of humanity itself: 'as smoke leaves
the fire never to return'.

• 69 •

When you press the button to shoot a zombie, you're pressing the button to make Mario jump. It's that simple.
   – Shinji Mikami

All games, whether played against machines or human opponents, are forms of behavioural software. The more sophisticated the programme involved, the less perceptible the separation between player and game. Illusion is consequently experience accelerated. Play against a slow machine, one that takes several seconds over each move, and you do not confront it directly. You know you're dealing with a process, a series of choices selected from a programmed list. Everything changes as soon as the machine starts playing as fast as, if not faster than, a human. At this point you have no option but to regard it as a real opponent. You must attribute beliefs and ambitions to it or lose the game. To use the technical term, the machine obliges its players to adopt the 'intentional stance', allowing them to make predictions about its future behaviour. In short, it replaces progression with narrative. The producer of *Resident Evil* is right to make the connection between taking a zombie down and Mario's jump. It is not the real or the definite that concerns us here but what is believable. Simulation can only take place in real time. Defined as a response that takes less than twenty to thirty milliseconds to occur, real time permits humans and machines to interact effectively with one another. Any slower than that, and delay becomes perceptible: the intentional stance disappears.

23, 81

Shooting a zombie in the head is a programmed reaction, entirely dependent on speed. Behind the basic game-play established by Shigeru Miyamoto in *Donkey Kong*, *Super Mario* and *Legend of Zelda*, out of which entire generations of videogames have developed, lie the comic books and animated cartoons of Osamu Tezuka. Just as many modern videogame designers regard Miyamoto as a founding god, so Tezuka-*sensei* is revered as the *'Manga no Kami'*, or god of manga. Such veneration only makes sense when it is taken literally. Comic strips, cartoons and videogames are expressions of animistic belief. They tilt and invert human agency, repeatedly committing art's original sin of pleasing and convincing while at the same time breaking all restrictions and boundaries. The term manga denotes 'irresponsible pictures' and is derived from a series of humorous sketches and drawings produced by the woodcut artist Hokusai between 1814 and 1834. Tezuka renders the modern manga still more irresponsible with the introduction of sequential panelling and speech balloons borrowed from American comics. He further revolutionises comic strip design by bringing a cinematic attention to detail, pacing, and perspective; through the rapid cutting together of such effects he develops an accelerated narrative that is no longer dependent upon language. His irresponsible pictures are soon capable of adapting themselves to an expanding range of media formats, interests and trends. Everything exists to end up as a comic book. Manga create an environment dominated by speed; or as Tezuka himself expresses it: 'We are living in the age of comics as air.'

This acceleration of perception is nowhere more evident than in the home, especially after Tezuka brings his most famous creation Astro Boy to life as an animated cartoon character on Japanese television in 1963. 25, 103 25, 86 A powerful robot child, Astro Boy emerges from the smoke wreathing his father's laboratory in the year 2003, only to be confronted with a more dynamic version of the Electronic Delirium dominating the 1960s. He inhabits a speeded-up world of TV screens and radio receivers, automated factories, superhighways, rockets and space stations, which quickly extends into the modern household, especially when the Astro Boy animated series is picked up by NBC for the US. Where Astro Boy leads, others quickly follow: his Americanised urging 'Go! Go! Go!' is soon echoed by 'Go, Speed Racer! Go!' As with 8 Man, Gigantor, Marine Boy and Wolf Boy Ken, speed is what makes it believable. Such cartoons have little to do with reassuring pleasure: as with Romero's *Dawn of the Dead* an apocalyptic takeover of our world of experience is depicted amid scenes of invasion, breakdown and violent chaos. The increased instability of the cartoon universe has subsequently been so modulated in the West that companies like Pixar and DreamWorks Animation are producing characters who can smile and frown at the same time.

• 70 •

You have one life left – and then the machine kicks your ass. Trapped inside a pre-recorded behaviourist universe, with no intentionality of our own, we attribute it to our games instead. 'They are so designed,' note Adorno and Horkheimer, 'that quickness, powers of observation, and experience are undeniably needed to apprehend them; yet sustained thought is out of the question if the spectator is not to miss the relentless rush of facts. Even though the effort required for his response is semi-automatic, no scope is left for the imagination.' Their analysis actually concerns the effect Hollywood movies have upon their audiences but can just as easily be applied to the intimate, vectored space of the videogame. An obsession with toys and gadgets often accompanies feelings of resentment towards existing power structures, offering the security of interactivity and ironic detachment. The videogame console, like the domestic tape deck, transforms the experience of life in a modern pushbutton home into something interactive. That is to say, it gives the illusion of control. Interactive experience in the home is never personal: leaving little room for self-reflection, it offers only a simulated ordering of subjective responses. In the programmed environment of the videogame, it escapes notice altogether. Reality becomes a game than runs itself. The more real our games become the more irony is our sole recourse, manifesting itself as a condition of informed stupidity. 'I never got the timing down with Mario Brothers, and never got the timing down to enjoy a normal calm relationship,' confesses one veteran player. Game over.

In the real-time regime of interactivity, playback has overtaken itself. Reactions make a videogame. Very few run in silence, however. Random noise and audiovisual scrambling of information are vital by-products of the gaming urge. As the games console slowly evolves into an all-encompassing home entertainment system, sound becomes a facilitator, creating greater involvement out of the digital rattling of empty shell cases. To fashion a weapon from the instrumentation of such distractions is to admit that music reflects an absence of individual power. 'Playability' is consequently a concept that links games and music to

63, 84

machines. 'Videogames run through my life like the music I listen to,' our veteran gamer recalls. In behavioural terms both can only simulate action; we conceal our true selves within them even as they operate through us. Connecting up hand to eye to brain, playability also helps position the body in what was formerly taken to be personal space. 'A painting by Richard Lindner, "Boy With Machine," shows a huge, pudgy, bloated boy working one of his little desiring-machines, after having hooked it up to a vast technical social machine,' write Guattari and Deleuze at the start of *Anti-Oedipus*. Pale and domesticated, happily connecting up personal to social space, Lindner's 'Boy with Machine' depicts the disciplined child who has already got his timing down. Game over.

Consciousness is manipulated through the senses, and our existing media serve to document the precise history of this relationship. A UK poster campaign for the game *Halo 2* shows a devastated burning landscape; an anonymous figure in body armour fills the foreground, features hidden behind a darkened visor, brandishing an automatic weapon. Accompanying this apocalyptic scene is a single statement: 'Movies Won't Matter Anymore.' It is an assertion that can be read on at least three separate levels – that *Halo 2* is so addictive that nothing else will be capable of claiming your attention; that the disaster depicted in the game renders such pastimes irrelevant; and that videogames are the apocalypse so far as the future of movies and other media entertainment is concerned. The discipline of simulated combat has become industry standard: invasion is the established means by which haptic space is entered and explored. The outskirts of ruined cities multiply themselves across advertisements on TV screens, magazine spreads and public hoardings. Darkened faceplates, protective armour and handheld weapons transform the player into an anonymous disciplined body, completing the simulation of the gaming experience. The marketing campaign for the perfectly-named *Crysis II* even goes so far as to urge potential players to 'Be The Weapon'. Game Over.

40, 87

• 71 •

Watch this protest in reverse!
Ask the local police; 'What's your illegal activity on duty?'.
If you protest the government then there's a new government from
protesting.
There's not a new government from protesting.
Thus, you aren't protesting the government.
6, 74   *– Jared Lee Loughner, extract from text accompanying YouTube post*

'We see from the foregoing,' notes Carl von Clausewitz, 'how much the objective nature of War makes it a calculation of probabilities; now there is only one single element still wanting to make it a game, and that element it certainly is not without: it is chance. There is no human affair which stands so constantly and so generally in close connection with chance as War.' Director of the Military Academy in Berlin between 1818 and 1830, Clausewitz attempts in his treatise *On War* to leave as little to chance as possible by systematically anticipating and listing the determinant factors and possible outcomes of any organised conflict. Beyond science and art, the waging of war constitutes an engagement with contending forces in which chance cannot help but play its part. War is therefore, Clausewitz argues, 'a wonderful trinity, composed of the original violence of its elements, hatred and animosity, which may be looked upon as blind instinct; of the play of probabilities and chance, which make it a free activity of the soul, and of the subordinate nature of a political instrument, by which it belongs purely to reason.'

By characterising conflict as the continuation of policy by other means, however, *On War* allows itself to be read as the handbook for a game in which the 'free activity of the soul' is constrained by very definite limits. Games, like the practice of magic, have both unseen outcomes and unforeseen consequences. What makes Mario jump is not necessarily what also makes the newsreader laugh – or to paraphrase William Burroughs: just how free is free? The relationship between the two conditions is encrypted within the politics of simulation itself: we continue to 'know so much that we don't consciously know'. During the late 1930s,

Johan Huizinga postulates the existence of a 'magic circle': a hypothetical barrier that separates those involved in playing a game from any awareness of the outside world. By conforming to the rules of the game, players limit their consciousness to it at as well: their actions are dictated by logic and rational choices that can be anticipated and pre-programmed. Huizinga's concept of *Homo Ludens,* or the 'play element of culture', is taken up aggressively in the 1950s and 1960s, when the principles of behavioural psychology increasingly influence everything from architectural design and the planning of advertising campaigns to the implementation of national defence and diplomatic strategy. Games affirm the running of individual programmed events that exist separately from the processes of history yet are framed by them. Otherwise they are considered to have no influence, thereby rendering them meaningless and without an accepted sense of order. None of the players are able to get their timing down. Jared Loughner's reversible logic has led him to stand trial for firing into a crowd outside an Austin supermarket, killing six, injuring fourteen and seriously wounding US Congresswoman Gabrielle Giffords in the head.

46, 82

'The founding idea of CNN is Event,' observes a former television news executive. 'Getting the signal out of there is our triumph... no matter that Event is no longer the primary interest of viewers.' Or to put it another way: Network News Won't Matter Anymore. Moving outside the magic circle of the game alters your relationship with it irrevocably. Burroughs recalls hearing 'the drunken newscaster': a random tape *'cutting up news broadcasts i can not remember the words at this distance but I do remember laughing until I fell out of a chair.'* News is the encoded simulation of an event imposed from outside that event. All games, rites and rituals depend upon codes to maintain their power. Reality is no longer something calculated as it was in the age of the computer mainframe – it has to be encoded as gameplay instead. Obscuring events by transforming them into a programmed list of behavioural options, gameplay in turn offers only a simulation of the random. 'You have an advantage which your opposing player does not have,' Burroughs instructs those interested in cutting up such codes. 'He must conceal his manipulations. You are under no such necessity.'

1, 80

· 72 ·

I may mention here that radio-aerograms are seldom if ever used in wartime, or for the transmission of secret dispatches at any time, for as often as one nation discovers a new cipher or invents a new instrument for wireless purposes its neighbours bend every effort until they are able to intercept and translate the messages. For so long a time has this gone on that practically every possibility of wireless communication has been exhausted and no nation dares transmit dispatches of importance in this way.

39, 108
                   – Edgar Rice Burroughs, The Gods of Mars

Code is the strategic organising of behaviour. Exposing the workings of rationality by running it out of sequence with itself, a code is therefore not a text but a process that resides outside of that text – *just* outside. EVP's cryptic messages drift through organised airspace. They constitute a random rejoinder to the strict regulation of broadcast media imposed 'from above' by which everyone receives a one-way mass communication to which they can never respond. EVP are the product of electronic media forced into conflict with themselves. Painstakingly elaborating a dialectic of chance in his treatise *On War,* Clausewitz systematically attempts to anticipate the one thing that will forever elude his text. What remains safely contained within every line, however, is the implementation of physical force itself.

'Simulated force is powerless,' declare Adorno and Horkheimer from behind the grey veil of World War Two. 'Culture has developed with the protection of the executioner.' When STRICOM, the Simulation, Training and Instrumentation Command for the US Army, is established in 1992, it operates under the motto 'All But War Is Simulation'. Or to use the marketing slogan adopted to sell the 'Commando Elite' of interactive war toys in Joe Dante's 1998 movie satire *Small Soldiers*: 'Everything Else Is Just A Toy'. Richard Lindner's 'Boy with Machine' assumes its full significance. The rise of the videogame also marks the resurgence of representational art; what Marcel Duchamp once dismissed as 'retinal painting' is reconfigured as a digital environment. The shift from analogue culture to the Digital Regime is marked by a corresponding shift from 'signal-to-noise' to 'figure and ground' as the main model for 44, 79 comprehension. A corresponding change in pathology occurs from intellection to immersion, perception to reaction, and calculation to encryption. Not that these conditions are in any sense mutually exclusive: Cocteau's description of his movie *The Testament of Orpheus* as 'a syntax of images instead of a story accompanied by words' could just as easily apply to the action of a videogame.

Cocteau and Burroughs both play with guns during their careers: poets rarely imagine a peaceful death for themselves. The characters in Cocteau's Orphic trilogy are constantly falling, pierced and penetrated, fatally wounded in the head. Burroughs fatally wounds his wife Joan Vollmer in the head recreating a sharp-shooting act straight out of *Gun Crazy*, a B-movie noir released the previous year. Cocteau and Burroughs also share a fascination with running irrevocable acts backwards and forwards in recorded time, removing any degree of personal commitment. Family money helps separate William Burroughs from the consequences of his act; while electing to cut up his texts calls into question his role both as author and actor. The dead always come back to life in Cocteau's movies: pistol shots propel the phantom poet through cinematic time in *The Testament of Orpheus*. 'Professor,' he complains at one point, 'it is difficult to explain the intertemporal and even more so to live in it. One gets confused.' No wonder ghosts express themselves so cryptically as EVP. Finally it is only death that fixes the poet in time. Shooting a zombie in the head is all that it takes to stop the dead from rising and walking again. Movies are disasters waiting to happen: it is written into them. More than a denial of death, running the film backwards and forwards is the denial of chance: it makes light of the possibility of failure. 'Watch this protest in reverse!' writes Jared Loughner. Congresswoman Giffords recovers from her head wound.

More than 147 years on from the time of its original writing, a coded message sent in a bottle to a Confederate general defending the town of Vicksburg Missouri during the American Civil War is finally deciphered. Inscribed onto paper only three years after Martinville uses his phonautograph to register 'Claire de Lune', the encrypted text is dated 4 July 1863: the same day that Vicksburg fell into Union hands. 'You can expect no help from this side of the river,' is all it says. No EVP has ever been more succinct. Instead of provoking laughter, however, this particular communication brings only a shiver of recognition. The Digital Regime expertly organises our perception of 'emergence': the stacking of simple instructions that leads to convincing but unexpected forms of agency. This is, after all, what connects figure to ground, allowing Pac Man, Mario and Sonic to run their respective labyrinths – and coded messages to become unscrambled into individual thoughts, words and behaviour.

61, 74

· 73 ·

Memory seems to be a no more stringent constraint than processing power. Moreover, since the maximum human sensory bandwidth is ~10^8 bits per second, simulating all sensory events incurs a negligible cost compared to simulating the cortical activity. We can therefore use the processing power required to simulate the central nervous system as an estimate of the total computational cost of simulating a human mind.
  – Nick Bostrom, 'Are You Living in a Computer Simulation?'

Hackers' fingers in rapid motion on a QWERTY keyboard before a flickering monitor have become a cinematic habit, indicating a form of magic that initiates remote effects and invisibly influences events. What permits the hacker and the videogame player such powerful dexterity is the expansion of the QWERTY arrangement through the addition of the shift, ctrl and alt keys. Together they make possible a typographical manipulation of text that remains just outside text: within this magic circle of encryption language is reversed and inverted in order to transmit words of power, which could otherwise not be read or spoken without being activated at the same time. The gamepad and the handset put us in touch with ghosts and spirits without our even noticing it anymore.

'Nothing is true. Everything is permitted,' declares Hassan-i Sabbah. 'Not actual gameplay footage' runs the standard disclaimer on television advertisements promoting videogames. Also known as 'The Old Man of the Mountain', Hassan-i Sabbah is remembered today for leading a breakaway Ismaeli sect of Shi-ite Muslims in the eleventh century dedicated to the overthrow of the Sunni Caliph of Baghdad and the end of Turkish rule in Persia. He is credited with the use of hashish to programme his assassins to kill – the term being derived from the Persian word *hashishim* meaning 'users of hashish'. Invisible brothers inspired by smoke-wreathed visions of paradise as their reward, Hassan-i Sabbah's assassins are adept at simulation, blending in with the outside world, sharing their victims' lives until the moment comes to strike.

3, 80

Invading present time in 2007, the videogame *Assassin's Creed* offers a fictionalised version of the Hassan-i Sabbah legend. An unseen assassin in a desert environment is sent out on a mission to kill. He exists, however, only as a set of experiences reconstructed by someone else as an immersive form of simulated gameplay located in the future. The real game lies in maintaining a degree of synchronisation between the player and the killer who went before him. During *Assassin's Creed* the screen even breaks up with random strings of code to remind players they are living out someone else's simulated experiences. That same year *Halo 3* is launched with a publicity campaign focussing on just one word in block capitals: BELIEVE. Movies won't matter anymore because we will have become the movie – emergence has replaced transcendence as our model for consciousness. Another inversion occurs in the history of human communication when the avatar, formerly regarded as the divine manifestation of the transcendental on the material plane, becomes instead the digital manifestation of the all-too human in cyberspace.

We 'dematerialise' by granting a greater significance to our coded online selves. The invisible generation play their online games like assassins: without drawing attention to themselves. In a classic reversal they have made cyberspace their lower plane. What, after all, is a videogame without its levels? According to Nick Bostrom, our lives may turn out to be programmes developed by a post-human society located in what we still regard as the future. 'If we learn more about posthuman motivations and resource constraints, maybe as a result of developing towards becoming posthumans ourselves,' he argues, 'then the hypothesis that we are simulated will come to have a much richer set of empirical implications.' The 'Eternal Return' re-emerges as the digital simulation of every detail that ever went before – 'even this spider and this moonlight between the trees', as recorded so obsessively by the likes of Erkki Kurenniemi and Gordon Bell. The future remains what happens after you're dead. Rather than worry about this prospect, however, Dr Bostrom suggests that we accept the situation and act normally. Play against the rules and you risk being deleted. Game over.

25

Requiem for the Network

'Consciousness is really just a net connecting
one person with another – only in this capacity
did it have to develop: the solitary and predatory person
would not have needed it.'

Friedrich Nietzsche
The Gay Science

· 74 ·

'Troy Town' is a historical place name denoting the former location of
a turf maze or labyrinth cut directly into the earth, the complex design   72, 87
of its pathways reflecting a need for security as mythical then as it is
absolute today. 'The tradition which accompanied the plan,' one
historian reports, 'is that the city of Troy was defended by seven walls   5
represented by the seven exterior lines and the entrance made as
intricate as possible in order to frustrate an attacking force.' Security has
remained a constant condition ever since; and yet we continue to lose
ourselves in it, afraid of what we might find there. 'The Enemy is
permanent,' observes Marcuse in *One-Dimensional Man*. 'He is   15, 59
not in the emergency situation but in the normal state of affairs.' And
he should know, having worked for the Office of Strategic Services
during World War Two. Responsible for 'the planning, development,
coordination and execution of the military programme for psychological warfare', the OSS was the forerunner of today's CIA. Numbered
among its alumni are radical philosopher Norman O Brown, the an-   33, 57
thropologist Gregory Bateson and Henry A Murray, cofounder of the
Department for Social Relations at Harvard. Brown later befriends John
Cage, while Bateson arranges for poet Allen Ginsberg to have his first
LSD experience at the Mental Research Institute in Palo Alto California,
and Henry A Murray subjects a gifted young mathematics student
named Ted Kaczynski to a series of psychological stress tests.

Conducted under laboratory conditions, Ginsberg's hallucinatory experience leads him to fear that he is about to be 'absorbed into the electrical network grid of the entire nation', while Ted Kaczynski, better known to the FBI as the 'Unabomber', goes off that grid altogether, releasing letter bombs into the system from his remote cabin in the woods. Solitude, according to Marcuse, has become technically impossible. However, the very condition that sustains the individual 'against and beyond his society' informs today's networked modes of existence. The network is, in effect, the hollowing out of a broader and deeper spiritual truth: that we are all linked by a shared sense of isolation and dependence. It is also through the network that the masses, that most modern and obsolescent of concepts, encounter and discover themselves. 'Now you can get four clicks into the organisation,' a Microsoft employee boasts to readers of *Wired* in 2007 concerning how easily outsiders can gain access to the company's main website. One month later *The Economist* publishes a story revealing that al-Qaeda is poised to infiltrate the windswept halls and haunted pavilions of *Second Life*. Such reports invariably identify al-Qaeda as a 'terror network'. The connection is not hard to make.

So who passes through the Net and who gets caught? The question is not as fanciful as it might at first appear. When the system works, you cannot get beyond it; and the mind of an outsider is even harder to comprehend when there is no longer any outside. 'The West was focussed not on the Iranian people but on the role of Western technology,' observes one Persian blogger, referring to mainstream media accounts of the role played by social networking sites in the anti-government protests of June 2009. The Enemy has become so deeply embedded in the system that the Enemy finally is the system. 'How did the unrest after the Iranian elections come about?' the *People's Daily* asks on behalf of the Chinese Communist Party. 'It was because online warfare launched by America, via YouTube video and Twitter microblogging, spread rumours, created splits, stirred up and sowed discord between the followers of conservative reformist factions.'

15

51, 82

6, 71

By expelling the outsider, the network internalises intrusion and isolation, making the Enemy an intimate part of itself. The prospect of everyone reading from the same Bible, unfolding the same newspaper or watching the same TV channel, with the Enemy positioned permanently outside of the system, is no longer a tenable state of affairs. Like can already speak with like: like transcends space and time simply by resembling like. Networks develop from their ability to connect *unlike* with like, extending our senses over distance. Marcuse's Enemy, it seems, has been programmed into them from the start. In other words, it is not the intricate arrangement of city walls overlooking the planes of Ilium that produces the siege, but sieges that eventually produce cities. One look at the expressions on the faces of Barack Obama and his White House team as they sit surrounded by laptops in 2011 watching the raid on Osama bin Laden's walled compound in Abbottabad will tell you as much.

· 75 ·

Technological rationality reveals its political character as it becomes the great vehicle of better domination, creating a truly totalitarian universe in which society and nature, mind and body are kept in a state of permanent mobilization for the defense of this universe.
    – *Herbert Marcuse,* One-Dimensional Man

In 1830 Joseph Henry demonstrates that networked communication is possible over great distances by using electromagnets to ring a bell   118
at the other end of a mile of copper wire. In 1834 Charles Babbage   97
conceives of his Analytical Engine, capable of processing any mathematical formula by means of selected number cards: 'the Engine,' he asserts, 'will always reject a wrong card by continually ringing a loud bell and stopping itself.' What Lewis Mumford describes as the 'iron   34, 55
discipline' of the monastery clock, its chiming bell 'synchronizing the actions of men', extends its influence into the nineteenth-century factory system and beyond even that – the possibilities seem endless.   55, 91

The 'Victorian search engine' remains a fantasy of the twenty-first century, however. Between Henry and Babbage falls the shadow of Carl von Clausewitz. Published in 1832, his *On War* betrays an understanding of men and machinery that runs deep enough to embrace their faults and inefficiencies. 'The military machine, the Army and all belonging to it,' its author declares, 'is in fact simple, and appears on this account easy to manage. But let us reflect that no part of it is in one piece, that it is composed entirely of individuals, each of which keeps up its own friction in all directions.' Friction, he argues, has a negative effect upon all calculation, 'which no man can imagine exactly who has not seen war'. It is the wrong card that abruptly halts the workings of the Analytical Engine and will later bring a modified Blackhawk helicopter down inside the Enemy's walled compound.

According to Clausewitz's reading of the term, 'friction' is an early expression of the relationship between signal and noise. *On War* ranks it alongside only 'information', which to the nineteenth-century military mind is more likely to be misleading than otherwise, as a disruptive factor in the outcome of any conflict. Friction, in other words, is the impedance of data. Precision of meaning, formality and constraint become characteristics of a purely analogue regime: the kinematic coupling of words and mechanisms is designed to reduce inefficiencies and misunderstandings. 'Our writing tools are also working on our thoughts,' Nietzsche will famously remark in a letter from Italy while learning to use an early model typewriter. But even the century's first mechanised philosopher despairs of its slowness and illegibility – especially when the Italian spring turns humid and the ribbon grows sticky, trapping the keys in ink and further impeding both writing and thought.

63, 96   Typographic sentences are organised in straight lines; but their content and meaning are capable of establishing complex relationships with other texts in the form of quotes, paraphrases, bibliographies, concordances and footnotes, meaning that every book has the potential to establish its own network of references. So when in 1842 the Italian military engineer Louis Menabrea completes his *Sketch of an Analytical Engine Invented by Charles Babbage* Ada Lovelace, Babbage's guide and mentor, immediately sets about translating it into English. The only daughter of Annabella and Lord Byron, it is Ada who has formulated the intricate mathematical formulae behind the 'Analytical Engine'. 'It must be evident,' she observes in her commentary to Menebrea's sketch, 'how multifarious and complicated are the considerations. There are frequently several distinct sets of effects going on simultaneously; all in a manner independent of each other, and yet to a greater or less degree exercising a mutual influence... All and everything,' she concludes, 'is naturally related and interconnected.'

Even as it articulates the shifting complexities of nineteenth-century calculus, Ada Lovelace's insight anticipates the unimpeded networks of the Digital Regime. It looks forward to a shift away from analogue's constant conflict between signal and noise towards the free flow of data. Such universal interdependence can exist only in the absence of Clausewitz's friction – otherwise the system loses all meaning instead of being flooded by it. As the network of typographical texts is extended by machines over distance, history becomes transformed into a series of headlines. 'What hath God wrought?' may have been Samuel Morse's first coded communication to be sent by wire from Washington DC to Baltimore in 1844, but it is followed by the anticlimactic inquiry: 'Have you any news?' This request is repeated in 1858 over the first transatlantic cable: 'Pray give us some news from New York, they are mad for news.' However valuable such messages may be, they are also expensive, requiring economy of expression. Accounted for word by word, sentences become shorter, stripping events down to their barest essentials. Language develops a new uniformity based on familiarity and superficiality; and meaning itself is regulated by the demands of transmission.

58, 86

65, 88

The telegram quickly debases the language of 'interconnectedness': further restricted by the subsequent distribution of the first telegraph offices along the existing network of railway lines. This association of steam powered engines with the receiving, storing and forwarding of messages suggests how dependent communication has become on industrial models of organisation. The rail and telegraph networks constitute huge factories processing information as rapidly as possible and over great distance. 'The railroad mania withdrew from other pursuits the most intellectual and skilful draftsmen,' Babbage complains after the development of his unfinished Analytic Engine has exhausted all its technical resources. In its place sits the lone telegraph operator in shirtsleeves and green eyeshade at his desk on the railway station, tapping out code: receiving, storing and passing on short messages that require urgent dispatch. Telegrams are unsurprisingly associated with bad news in the popular imagination. Misunderstandings and paranoia set in; alien threats start to appear. 'Imagine a gigantic telegraph network encompassing the universe and converging on a single centre,' runs an anti-Jesuit tract from 1870, 'each member of the Society of Jesus is a wire, the General is the center.' Scandalised by its homoerotic overtones, *The Scots Observer* accuses Oscar Wilde of writing *The Picture of Dorian Gray* for 'none but outlawed noblemen and perverted telegraph boys'. As the network spreads and grows, the Extension of Man increasingly becomes the Enemy.

• 76 •

Implicit in the use of words (when messages are put across) are training, government, enforcement and finally the military. Thoreau said that hearing a sentence he heard marching feet. Syntax, N. O. Brown told me, is the arrangement of the army.
        – John Cage, 'The Future of Music'

'This has not been a scientist's war,' Vannevar Bush declares in the summer of 1945. 'It has been a war in which all have had a part.' The fact that the head of wartime scientific research feels the need to make such a disclaimer speaks volumes. It comes at the start of 'As We May Think', in which he ponders how the 'spider web of metal' within each thermionic valve will enhance the storage and transmission of data during peacetime. In the 'cheap complex devices' Bush describes so enthusiastically in this influential essay, the network is to find its most responsive

65, 89

54, 94

48, 54

48, 83

medium – and among the scientists whose activities he has coordinated as part of America's war effort, some of its most willing adopters. A
new generation of sophisticated machines starts to emerge from low-
cost, mass-produced components, demanding unprecedented levels
of technical expertise. 'Babbage, even with remarkably generous support for his time,' Bush boasts, 'could not produce his great arithmetical
machine.' Through the increasing interconnectedness of electronic
parts, however, the metallic spider's web will soon be expanded into a
labyrinth of such machines – even though, as Bush himself admits, 'the
users of advanced methods of manipulating data are a very small part
of the population'.

Converted by advanced data manipulation into a repetitive processing of facts and statistics, the masses will henceforth exist only as a
set of projections and outcomes. To this end, everything processed
is accorded an equal value; and individual survival becomes the
sole grounds for human conduct. The free play of information as an
economic commodity will no longer respond to mathematical laws
alone, however. 'If scientific reasoning were limited to the logical processes of arithmetic,' Bush points out, 'we should not get far in our understanding of the physical world. One might as well attempt to grasp
the game of poker entirely by the use of the mathematics of probability.'
Survival is thus transformed into a game played against an opponent
who remains only dimly perceived and yet constantly present: the
product of a stunted awareness. From the wartime development of
sonar and radar systems to the increased receptivity of electronic
media in peacetime, the nervous probing of hitherto unsuspect-
ed spaces for signs of the Enemy has been subject not to technical
limitations but ideological ones. To extend awareness to the point
where we experience the pain of others may ultimately prevent us from
harming anyone. Sensitivity and detection must therefore be placed
in expert hands. Poker-faced and fastidiously removed from all direct
experience, such specialists inhabit a deepening sense of isolation; the
war in which 'all have had a part' is to be run by an elite for whom, as
Clausewitz puts it, 'science must become art'. As the balance of military
power shifts from those with the biggest army to those with the most

advanced weapons technology, the tactical demands of waging war with an unseen enemy grow alarmingly. Art as ritual is replaced by science as ritual, the outcome determining who wages war and who will merely endure its consequences.

52   Just before the outbreak of World War Two, Le Corbusier designs 'a museum of unlimited knowledge' by flattening a pyramidal spiral onto a horizontal plane, creating 'a labyrinth with fluid walls in which all shortcuts and detours were allowed'. In 1953, only eight years after the publication of 'As We May Think', the RAND Corporation moves into its new Santa Monica headquarters. Arranged in a two-storey grid around a set of square courtyards, the building has been specifically designed by mathematician John Williams, using these same 'shortcuts and detours', to maximise random encounters between RAND's various specialists in Cold War strategising, thus preventing them from grouping together within their respective disciplines. Similarly, Norbert Wiener's daily wanderings around the campus at MIT are facilitated by the institute's unique layout, incorporating the repeated contortions of a single 'infinite corridor' which opens onto identical clusters of offices and classrooms. 'It has twelve covered courts with opposite doors,' Herodotus writes of the Great Labyrinth built to commemorate the rulers of Lower Egypt, 'six courts on the North side and six on the South, all communicating with one another and with one wall surrounding them all.' Meanwhile, down in the 'labyrinthine basement – somewhere under the Snack Bar' at RAND HQ, Marcuse notes that 'the world becomes a map, missiles merely symbols and wars just plans and calculations written down on paper'. This is the tactical nuclear 'war game' played
52, 92   out as a cynical form of theatre beyond anything Nietzsche, Wagner or
24, 43   Brecht could possibly imagine. As opposing Blue and Red teams run every possible variation on the same mutually-destructive scenario, the only option not open to all participants, Marcuse wryly points out, is 'negotiation'.

• 77 •

'All action in War,' Clausewitz asserts, 'is directed on probable, not on certain, results.' Mathematically precise models for determining rational strategies in the face of uncertainty have been developed by John von Neumann, a consultant to the Los Alamos National Laboratory, where the first atomic weapons were created. Known informally as 'game theory', these offer a quantitative approach to estimating the potential outcome of any conflict. Wiring strategists and soldiers, mathematicians and logicians into the decision-making process, game theory assumes that the Enemy will always be rational in his behaviour. This rationality, however, is based entirely upon the notion of compulsion: that the Enemy has no choice but to do whatever he is capable of doing. In fact, he cannot prevent himself from responding in the most devastating manner possible. Outcome now supersedes all other factors. Physically remote from the local conditions of any actual conflict, the blast-proof command bunker – whether in the form of a remote radar station or a secure war-gaming room – is the rational expression of an Enemy seen only as a set of 'rational' compulsions.

To minimise uncertainty is to establish the networked flow of data as a form of conflict without friction. The network consequently regulates what RAND analyst Herman Kahn is pleased to call 'more reasonable forms of using violence'. This new order of military theatre, drained of all passion and destructive urges, is the drama of the Operant Conditioning Chamber designed by behavioural psychologist B F Skinner. A quantitative approach to the unpredictable complexities of human psychology, behaviourism assesses its subject as little more than a series of responses to external stimuli. It is the Cold War mindset reduced to a single seemingly abstract principle: one that enables powers in the East and West to sequence the destruction of their own people into a timetable of reciprocal strikes. The ground to all game theory, operations research, communication theory and systems analysis, behaviourism sets the rat down in the laboratory maze and pits the Red team against the Blue. It also enables humans and machines to interact at unprecedented levels of intimacy.

23

62, 87

93

19, 105
25, 118

63, 101

In response to the launch of Sputnik I by the Soviet Union, the United States sets up the Advance Research Projects Agency. Otherwise known as the ARPA, it fosters innovative approaches to scientific thinking by funding research through a growing network of military, industrial and academic organisations. At the same time, the United States Air Force, in association with IBM, Western Electric, Bell Telephone and MIT's Lincoln Laboratories, comes up with SAGE, or the Semi-Automatic Ground Environment. A gigantic air defence computer built into the basement of a three-storey blockhouse, SAGE can be accessed only through keyboards, screens and light-guns, allowing its operators to call up and process information derived from a network of radar installations ringing the outer perimeter of the United States. SAGE is the first system to deploy digitised data over telephone lines and to present complex tides of information via graphic displays. A former Gestalt psychologist whose own experiments echo those of B F Skinner, Joseph C R Licklider of MIT is in part responsible for structuring this new relationship between computer and human, which he broadly categorises as 'symbiotic'.

· 78 ·

We tried to get AT&T to build the network. The Air Force was prepared to pay all the costs of building it. We did studies of the cost, and we figured it would be 60 million dollars a year to build the network and maintain it. The government was paying somewhere between one and two billion dollars a year to AT&T for long-distance communications. So we figured it could be a bargain. AT&T, which was the telephone monopoly at the time, was dead set against. First they said it couldn't possibly work. They were violently opposed to the idea. Sometimes it was hard to recognise who the real enemy was...
     – Paul Baran

'It is the acts of men that survive the centuries, which gradually and logically destroy them,' declares the supercomputer at the heart of Jean-

Luc Godard's Alphaville. 'I, Alpha 60, am simply the logical means of   10, 90
this destruction.' During the Electronic Delirium, *On War* reads less like   68, 102
a playbook on human conflict and more an operator's manual for Planet
Earth. Strategies are now being put in place for what will effectively be
the first and last truly global event – and one that seems rational only
as a series of theoretical propositions. Herman Kahn calls it 'thinking
the unthinkable', while Buckminster Fuller sees it as defining the choice   55, 113
between Utopia and Oblivion. Newly arrived at RAND's Computing   59, 88
Division, Paul Baran contemplates the problem of USAF requirements
for 'minimal essential communications' following an enemy nuclear
strike. Maintaining command and control of nuclear weapons following a pre-emptive attack requires a different kind of communications
network: one that is neither as rigidly organised as the telephone system nor as susceptible to the electromagnetic pulse generated by a
thermonuclear blast.

Baran's solution is to replace the standard central switch, through which
all telecommunications must pass, with what he calls a 'distributed net-   64, 93
work'. This is a loose arrangement of interconnecting nodes, each with
three or four separate links, which allows messages to follow different
paths to their final destination. He also incorporates the idea of 'pack-
et switching', to use the term derived for it by British physicist Donald
Davies, in which messages are broken down into individual bits of data
that are then allowed to flood the entire network. A receiving computer
then reassembles them into a readable form at the other end according
to a 'check sum' which determines that they have arrived in the correct
order. For each bit to pass through the system effectively, however, it
has to be digitised. Analogue data will not survive the process of being
received, stored and passed along that many times over the nodes of
a distributed network: with the accumulated deterioration produced
from each copy, the relationship between signal and noise is systematically inverted. The digital inundation of the distributed network, unimpeded by such friction, is free to overflow all limits: there will no longer
be an end to information exchanges of any kind.

Baran's new communications model differs so radically from the one maintained by AT&T that the company refuses to even consider it. 'The folks at AT&T headquarters always chose to believe their actions were in the best interest of the "network",' Baran later recalls, 'which was by their definition the same as what was best for the country.' Encountered deep within the system, the Enemy is easily mistaken for a rival service. Even so, RAND makes sure that information about Baran's model is available in Russian in the hope that the Soviets will improve their own minimal essential communications: at this level of interactivity neither side can afford for the other to make a mistake. Seemingly connected up only to itself, the telephone system has left itself open to constant Enemy incursion. When AT&T refuses to grant access to its Long Line maps, RAND engineers are obliged to study the network's vulnerabilities from a set of stolen plans. Meanwhile, as the shadows start to lengthen across the 1960s, an anonymous online community of youths, armed only with a telephone handset and a few simple operational principles, have also figured out how to get inside AT&T's wiring, hacking into the network as if it were a gigantic computer. When not setting up illegal chat rooms in major communication hubs, dialling long-distance for free, or attempting to prank-call everyone from the Pope to the President, some of these more gifted and ambitious 'phone phreaks', if they do not wind up in jail, go on to found their own IT companies.

· 79 ·

'Is there a difference between the mystery of the laws of knowledge and those of love?'
— *Alpha 60*, Alphaville

Everything can now be copied, stored and transmitted without end. The old analogue relationship between signal and noise has been replaced by a digital patterning of zeros and ones which more closely resembles the one between figure and ground described in Gestalt psychology. Individual perceptions establish relevance and form: meaning

72, 118

is created by a degree of interactivity with data that is as much psychological as it is mathematical. Just as immediate experience must form a coherent whole, so the network must be complete. Or as John Heider of Esalen puts it, stating the same proposition in terms that Ada Lovelace would understand: 'when the figure is present and the ground fraught with energy, all creation reveals meaningful interconnectedness.' Figure and ground estalish a dynamic relationship in which both are simultaneously separate and inextricable from each other. A network of neon lights, flashing texts and signs, Alpha 60 expresses the integrated awareness of an entire city; and yet Godard's *Alphaville* presents this consciousness as functioning without 'conscience' or 'tenderness', to identify two qualities that the film's protagonist Lemmy Caution claims are lacking from the population. Extending to others our sensitivity to pain opens up an erotic dimension to interconnectivity. Sex and death inevitably form part of the equation: the networked mind encounters the 'ground fraught with energy' as a profound form of hallucination.

64, 106
63, 98
66, 91

Ecstasy is force that has grown aware of itself; beyond cause and effect, it breaks with all behavioural constraint. Is there any wonder that those who have experimented with hallucinogens are quick to grasp the revolutionary potential of the computer? A spiritual 'think-tank' established on the California coast at Big Sur, Esalen effectively hardwires tantric yoga and psychiatric theory, cybernetics and psychedelic experience together. This interdisciplinary fusing of Eastern and Western thought is captured on the cover of its earliest publications, where a line drawing of a lotus is juxtaposed with a mathematical formula marked 'infinitesimal calculus, Bertrand Russell'. Even the interconnecting system of hot springs, saunas and massage rooms for which Esalen is famous appear to have been connected together in a neurological manner. Teachers at the institute include Gregory Bateson, B F Skinner, Buckminster Fuller, Dr Timothy Leary, and the father of Gestalt Therapy, Fritz Perls.

What separates Utopia from Oblivion now is a distributed network of digital services. Former Gestalt psychologist Joseph Licklider is appointed head not only of ARPA's Command and Control Office but also the Behavioural Science Division. The resulting new department, renamed the Information Processing Techniques Office (IPTO), continues to explore the future of man-machine symbiosis. Following Licklider's initiative, successive IPTO directors initiate the development of the ARPANET, the first distributed network of computers across North West America. A dynamic routing system of processors will relay data from one computer to another, hooking up RAND with Stanford, UCLA with Harvard and Bell Telephone with MIT. First, however, the processors have to be carefully 'battle-proofed' by welding them into steel safes before installation. It is even harder to recognise the Enemy this time around: rioting student protestors or the growing number of hackers eager to take a processor apart to see how it works. ARPA later 'battle-proofs' itself: adding a 'D' for Defence to its acronym. As DARPA, it will fund only weapons-related research, thereby releasing a flood of talented engineers and academics into companies such as PARC, Xerox and IBM, taking their state-funded expertise with them.

Simultaneously all figure and all ground, the Digital Regime represents the apotheosis of technology as environment. Such an acute awareness of the grid, however, can take the form of a disturbing and violent hallucination. While at the RAND Corporation, Herman Kahn experiments with LSD in the office of a Los Angeles psychiatrist, reviewing bombing strategies against China while under its influence. After leaving RAND for the Hudson Institute, Herman Kahn is a regular visitor at Tim Leary's Castalia Foundation, based at the Millbrook Estate in upstate New York and named after the intellectual community depicted in Herman Hesse's novel *The Glass Bead Game*. What the author of *Thinking the Unthinkable* has to say to the man who wrote *The Politics of Ecstasy* is anybody's guess. A former Harvard professor of psychiatry famous for his experiments with LSD, Dr Timothy Leary soon develops an interest in cybernetic programming. 'Tim will be the first person to get computers made illegal,' Esalen cofounder Richard Price remarks when told about Leary's enthusiasm for the Internet. Meanwhile John Cage notes how well the *I Ching* responds to being computerised.

3, 95

· 80 ·

It is the exact fulfilment of silent supposition, it is the noiseless harmony of the whole action which we should admire and which only makes itself known in the total result.
        – Carl von Clausewitz, On War

More eagerly anticipated than any before it, the twenty-first century begins ahead of schedule on the evening of 16 January 1991 at the precise moment when CNN reporters start broadcasting live from a room in the Baghdad Hilton, giving their first-person account of the bombing taking place outside the window. On 15 January, before American forces even go into battle, electronic warfare has already broken out. CBS, quoting an Israeli monitoring source, claims that the US is jamming Iraqi radio communications, while CNN is widely reported to be Saddam Hussein's main source of information on the West.

The global financial markets are expected to withstand the first twenty-four hours of military action in the Gulf, already predicted by one Pentagon official to be the most violent in the history of warfare. This is due to the series of checks and balances introduced into the system to prevent dangerously unstable fluctuations in trading following the World Stock Market Crash of 19 October 1987: a 500-point drop in value that prompted one White House Staffer to point out to the public that the stock market is 'no longer a place but a network' constantly shifting coordinates from one international time zone to another. People working in the US Federal Reserve on 16 January are consequently among the first to know about the start of Operation Desert Storm when the impact of the conflict on the US economy appears on their computer displays. They sit and watch the effects from minute to minute. At another American location, far from the actual Kuwaiti Theatre of Operations, personnel in the AT&T Network Operations Centre in New York are following the outbreak of the war on the wall-sized electronic map depicting their worldwide telecommunications systems. The circuits to Baghdad suddenly go dead at 7.30pm, Eastern Standard Time. Then the map lights up with record calling volumes to Saudi Arabia.

1, 71
3, 73

65, 93

25, 94

'It would not be logical to prevent superior beings from attacking the other parts of the Galaxies,' declares Alpha 60. Rolled out over military lines of command and control, cable news reports and financial indexes, Desert Storm will be a twenty-four-hour war for a twenty-four-hour economy. 'This is what our missiles see,' a young officer explains at an Air Force press briefing during Operation Desert Storm, running yet another mission video of exploding bunkers and wrecked supply columns. Detection and invisibility – the ability to see without being seen – is all about networked distance. 'Thus a future president of the United States might easily have command and control systems that involve having many television cameras in a future "Vietnam" or domestic trouble spot. Since he would be likely to have multiple screens, he would be able to scan many TV cameras simultaneously,' Herman Kahn suggests in *The Year 2000*, published back in 1967. 'Paul Nitze, the present Secretary of the Navy, has made the suggestion that similar capabilities might be made available to the public media or even the public directly. This would certainly be a mixed blessing. Obviously such capability can give a misleading psychological impression of a greater awareness, knowledge and sensitivity to local conditions and issues than really exists. This could lead to an excessive degree of central control.' The world is thus transformed into one vast Operations Room. Within the first twenty-four hours of Desert Storm advertising rates at CNN rise by over 300 per cent.

64, 90

· 81 ·

We must study carefully the ways of large industry, so that we can implement the fact that there is no limit to the place in which we live.
    – John Cage, 'Foreword', X

Being interconnected means that we are all ground – where then is the figure? By losing perspective along with our awareness of distance, we no longer share any basis for action: in other words, we lose a common sense of purpose. Information overload, according to IBM's old

formula, leads to pattern recognition. What connects Utopia with
Oblivion, it seems, is a distributed network of digital services. A woman
on a commuter train hunches in silence over her cell phone, listening   68, 107
intently – finally she says 'It's so tough when you get a taste of what it
can be like' and then falls silent again. The hollow truth is that we are not
connected to each other; everyone is connected to *someone else*. The
same electronic display at AT&T which tracked the first few seconds of
Operation Desert Storm also registers the system's collapse as phone
calls flood the network following the attacks on the World Trade Centre
and the Pentagon on 11 September 2001. In the Digital Regime pattern
recognition is transformed into a state of permanent crisis.

Distributed networks require saturation in order to function properly;
digital services thrive on excess. The seemingly limitless capacity
of free online storage, housing everything from spreadsheets,
maps and schedules to photographs, music and documents, is just
one by-product of this collective information overload. When did
you last delete anything purely for reasons of space? The network
subsumes so many areas of human activity – commercial, vernacular
and social – that its presence is impossible to read economically
except in terms of extremes. Free services are rated
by their numerical accumulation of individual users,
while IPOs and start-ups can run into millions of dollars, none of which
these same individual users will ever see. When did emails, instant
messaging and uploading data stop being work? When did spending your money online become a game? The range of free activities
that traditionally enhance life are reconstituted as a form of 'digital
folklore', while the social network replaces family and friends, taste
and consumption with a continually shifting set of connections.
Structure becomes subordinate to connectivity, and strategy to more
flexible forms of interaction. Assemblages of people and services,
images and impulses, none of which can exist in isolation from the
others, take shape in real time; by linking them together the network   69, 104
distracts us from noticing its presence. The question of who – or
what – gets connected is nowhere near as important as the point of
connection itself; and that can only renew itself from moment to

moment. 'Searching for Network' remains a standard screen message running on mobile devices.

On 15 January 1990, precisely one year before electronic warfare breaks out in advance of Operation Desert Storm, teenage hacker collective Masters of Deception crash the AT&T network while skirmishing with online rivals Legion of Doom. An estimated seventy-five million calls are blocked. Plugging their modems into the same system through which the 'phone phreaks' ran wild, youths like 'Acid Phreak' and 'Phiber Optik' are among the first to establish their own free-market reality within which individuals define themselves not by what they produce or consume but by the services at their immediate disposal. In the meantime those who have no business being inside the network are treated as criminals and made to behave accordingly. Images of the Enemy start to proliferate. The Masters of Deception, it turns out, operate from adolescent bedrooms in Queens and Bedford-Stuyvesant; while one member, as if to emphasise the importance of illusion, runs with the Decepticons, a Brooklyn street gang taking their name from the *Transformers* cartoon and toy line. By 1998 the editor of *Upside*, a technology magazine for venture capitalists, predicts that he will 'see the overthrow of the US government in my lifetime' – not by international terrorists but by Microsoft.

• 82 •

They have no definite approaches but wander about in circular side-tracks, and most savage monsters are concealed in their labyrinth by deception.
  – Henry, Abbot of Clairvaux

Before 1992, users of the Internet are required to sign a declaration that they will not conduct business over the network. No online commerce would therefore be expected during the saturation media coverage of Desert Storm. Today any number can play. The masses are a self-

71, 86

defining network occupying digital space. As they click distractedly in and out of the US Invasion of Iraq in 2003, issues of security and
transparency become increasingly posed in globalised commercial
terms. That summer Linden Lab launches *Second Life*: an online virtual
world in which customers are encouraged to run wild: you can be who
you want, build what you want, wear what you want, create an avatar
for yourself to have sex, make things and play with others. At the same
time the company inadvertently sparks off a revolt when they attempt
to charge *Second Life* residents for any artefacts they may create there.
Sporting historical costumes and slogans derived from the American
Revolution, the protestors organise themselves into the first flash mob
organised by Adam Smith, after which Linden Lab agrees that its citizens can retain their intellectual property rights.

In our collective rush to generalise we become bad at history. *Second
Life*'s largely forgotten sprawl may constitute some lost Atlantis today,
but its taxpayers' revolt should always be remembered. Prior to this,
online acts of enclosure converted areas of the Internet into a series
of medieval theme parks as simulated massive-world RPG communities generate economic growth from medieval systems of scarcity and servitude. A log-in identity gave instant access to the Middle
Ages in the form of Electronic Arts' *Ultima*, Sony's *Everquest*, Mystic's
Dark Age of Camelot and *Asheron's Call* from Microsoft. The Linden   4, 74
Dollar, however, fosters a hedonistic division of labour in which
bourgeois values are idealised. The trade in virtual goods and
services becomes correspondingly easier. If nothing else, this roll-
ing-out of laissez-faire economic liberalism over the Internet has giv-
en everybody on the planet the opportunity to experience crime for
themselves. The Office of Homeland Security establishes a branch in
Second Life amid heightened media reports of how terror networks
are using the site to recruit new members, plan future operations
and transfer funds. During the economic crash of 2008 news reports
describe lines of *Second Life* inhabitants forming at virtual ATM
machines to withdraw their Linden Dollars. Two years later Linden Lab
announces that it will lay off thirty per cent of its staff.

· 83 ·

Virus (11:01:29 PM): there is no 'love' on the internet
– IM exchange with 'Sabu', 11 August 2011

Security without territory supplies us with an Enemy but no actual war: it becomes a form of brand management instead. China has already accused the West of waging online warfare via YouTube and Twitter, while America holds the Chinese People's Liberation Army responsible for computer network attacks on its state department. 'In the past innovation has driven the military and corporate markets,' runs a business report from San Francisco. 'But now the consumer market, with its vast economy of scale and appetite for novelty leads the way.' Vannevar Bush's cheap complex devices have been replaced by cheap online services. The art of war now involves attacking systems and stealing sensitive data; botnets, worms and other malware have transformed human conflict into an online multiplayer game. The struggle, as always, is over access – and access remains primarily a question of identity. How else do you penetrate the seven walls of Troy except inside a Trojan Horse? 'Once you know how a weapon works in cyberspace, it can cease to become a weapon,' remarks Martin Libicki, a RAND digital warfare expert. 'The best weapon is one an enemy never knows exists.' In terms of brand management, the Enemy is an undetected virus.

For the individual user, the network has only one centre, which is the point of initial entry; all possible connections, the distribution of links and nodes, emanate from here. Each digital trail originating from this floating point constitutes an online profile of a consumer, whether of products, services, political opinions or social scandals. Digital data is automatically copied as it floods the network, forming what a Facebook executive has described as a 'foundational narrative timeline' for everyone. All and everything, in Ada Lovelace's words, is naturally related and interconnected through the multiple and repeated clicking of the 'Like' button, which constitutes a new libidinal economy of scale. A bodiless sublimation of Eros, such obsessive activity marks out individual lines of least resistance: we return again and again to the same pleasures

48, 76

and fascinations, generating endless copies as we express our online approval of them. Obsession ultimately becomes predictive as data is structured into a specific profile. With intelligence and business analysts studying futures markets as indicators of trends and events, DARPA proposes that an online gambling site use market forces to predict political assassinations and terrorist attacks in the Middle East. 'Total Information Awareness' provokes a political scandal, however, and its funding is cut off.   41, 97

• 84 •

Whatever Scrawlings are made upon the Walls of the Labyrinth by Travellers, these Simpletons swallow down for Prophecies   44, 87
   – G P de Tournefort, 1717

'Hello, my name is Jared Lee Loughner,' begins a YouTube post composed before the shooting of Congresswoman Gabrielle Giffords. 'This video is my introduction to you!' To the extent that any conspiracy is the unseen application of power, and all conspiracy theories are an attempt to disclose such practices, the Internet expresses both these compulsions perfectly. 'My favorite activity is conscience dreaming; the greatest inspiration for my political business information,' Loughner continues. 'Some of you don't dream – sadly.' Blogs, wikis and social network sites help to structure the world's data flows into a creeping form of awareness. As William Burroughs observes, the paranoid mind   63, 102
is one in full possession of the facts. While opposing ends of America's ideological spectrum blame each other's media rhetoric for his actions, Loughner's 'conscience dreaming' remains largely unexamined. What shaped his 'favorite activity' and helped to establish its connection with 'political business information'? What finally separates those who dream from those who 'sadly' do not?   2, 114

'He did not watch TV, he disliked the news, he didn't listen to political radio, he didn't take sides, he wasn't on the Left, he wasn't on the Right,' one of Loughner's high-school friends tells ABC News. Instead, reports suggest that he preferred conspiracy-based websites and the online musings of anti-government activists. Considered as an online state of mind, the Internet is a highly alert, constantly vigilant schizophrenic ceaselessly creating connections where none had existed before. Like Loughner's 'conscience dreaming', it functions best in a condition of hypnagogic lucidity, somewhere between sleeping and waking. Never has encroaching paranoia been so user friendly. What distinguishes Loughner's own diatribes is the degree to which he entrusted them to such open, sociably constructed platforms as MySpace and YouTube, where reality is whatever you make it: names and faces are interchangeable, and a 'friend' is just a thumbnail to be removed or replaced. Such connections are ultimately schizophrenic and therefore not detectable because they cannot easily be read.

We are no longer dealing with metaphors when it comes to conspiracies but with literal meanings and interpretations. In 2008 a policy paper from the director of the White House Office of Information and Regulatory Affairs proposes that government agents conduct a programme of 'cognitive infiltration' targeting conspiracy websites in order to undermine 'the crippled epistemology of believers' by planting doubts about their pet theories. While paranoia may well be the consequence of fully possessing the facts, Adorno and Horkheimer consider it the symptom of a 'half-educated' mind. They also diagnose the type of crippled epistemology that allows it to flourish; culture becomes a commodity disseminated as information that fails to have any effect upon the individuals distributing it. Thought is consequently restricted to the acquisition of isolated facts separated from any conceptual relationship that may have given them meaning. The Internet has always been its own greatest conspiracy.

40, 70

64, 107

· 85 ·

(9:03:29 PM) ZJ: ...it's hard to ensure security when the source is unavailable
(9:03:53 PM) bradass87: yes, even worse its often lowest bidder...
(9:05:03 PM) bradass87: used to be the cream of the crop... now its outdated non-backward compatible suites of buggy software that were originally used for civilian purposes, then modified for military but not exactly thoroughly tested
(9:05:36 PM) bradass87: then they get contractors who dont know anything about computers to teach it...
(9:06:42 PM) bradass87: and its all OKAY, because we cant exactly complain out in the open because the software which bugs out is often times on machines which are stamped with big red SECRET stickers
    – IM exchange between 'Zinnia Jones' and Bradley Manning, 21 February 2009

In December 2010 hackers collective Anonymous launches 'Operation Payback' against a number of online financial services, including Paypal and Mastercard, in retaliation for severing their links with WikiLeaks. A Distributed Denial of Service netbot program, a malicious piece of software designed to overload a website with more traffic than it can support, affects the targeted sites, either slowing them down or stopping them working altogether. Only in political terms is it possible to inconvenience your opponent into submission, and only in the Digital Regime can 'Denial of Service' be achieved through the flooding of the system. This is more than the online equivalent of picketing a store to protest its policies: the withdrawal of online presence is replaced by an exaggerated overemphasis on consumer activity. Think of this flooding in terms of lost customers, and it soon becomes clear that Distributed Denial of Service commits the unpardonable social crime of messing with the supply chain.

63, 92

11

Anonymity is an untraceable presence within the network. Individually and collectively, the anonymous hacker is all ground and no figure. No one is permitted to speak of it with authority without also arousing suspicion. Refusing to offer any personal data at all becomes an invasive practice that challenges everyone involved. Violence, which is a turbulent restatement of identity, emerges at the point of contact between the untraceable network presence and the physical world. *'All your base are belong to us'*: this online viral phrase expresses the connection in the most precise terms. Accounts of Bradley Manning's experiences as an intel analyst, remotely monitoring enemy targets from his workstation in Baghdad, highlight his dissociative behaviour: 'his mind in one place and his body in another'. From 'all ground' to 'all your base': what happens when social network mood-swings are replaced by pan-gender, ethnicity or employment changes? Giving your real name is not always possible or desirable even in those systems where it has become mandatory. Violence and identity, as Bradley Manning has discovered, are both forms of trespass.

Nobody wants free speech online if it means granting territory to people whose opinions or behaviour don't correspond to your own. Equally nobody wants free trade online: not when the segregated economics of abundance and scarcity can still be put to feudal effect. Post-industrial is beginning to look pre-industrial. 'Cloud computing' may shift processing power out of individual computers and reposition it online, but the giant data centres involved require huge amounts of hydro-electric power to cool the banks of processors employed by this floating system. The new citadels of the future, data centres are being built by rivers and aqueducts on the sites of old foundries and smelting works. Just like the great cities of the past, the data centre must be kept secure at all times. Even the Ancient Greeks, gathered on the plains outside Troy, would understand that. When one of its data centres burned to the ground in 2006, Google made a sales feature out of the fact that the fire did little to compromise the safety of the spreadsheets, documents and files it contained as these had already been relocated to yet another data centre. 'I have a staff of about 30 people dedicated to security,' marvels a head of IT at one large university. 'Google has an army; all of their business fails if they are unable to preserve security.' Meanwhile the Internet is transforming itself into an unfolding series of individual labyrinths as online services aggressively compete with each

17

other. The US military, instrumental in developing the ARPANET, has withdrawn into its own secret labyrinth, the Secret Internet Protocol Router Network or SIPRNet: a 'completely secure environment' for the transmission of classified data, where Bradley Manning once worked. 'It shifts from being a kinetic battle to siege warfare,' comments computer security expert Gunter Ollmann on how the threat of cyber warfare has seen many online security firms disappear from the Internet. 'This stuff is more kinetic than nuclear weapons,' adds Dave Aitel, yet another security expert. 'Nothing says you've lost like a starving city.' Here once stood Troy. *All your base are belong to us.*

Godzilla Has Left the Building,
And How He Got There

'...we would not like to be without the harrowing poetry
of their proximity, for these monsters often arrive when our
life, involved as it is in the spider's web of purposes,
has become too tedious and too filled with anxiety,
and provide us with a sublime diversion by for once
breaking the web – not that these irrational creatures
would do so intentionally!'

Friedrich Nietzsche
Daybreak

· 86·

Godzilla has been good to us over the years, the way only a bad dream can be. Monsters serve as warnings and should be welcomed as such. Having no place either in nature or in polite society, they have little choice but to exist as reminders of the things we would sooner forget. On 4 May 1897 over a hundred women and children are burned alive at the annual Bazar de la Charité in Paris when the flimsy replica of the capital's medieval *quartier*, constructed from painted cardboard and wood as a setting for the event, unexpectedly bursts into flames. The fire starts in the canvas booth set aside for the screening of a Lumière Brothers film show when the ether lamp used to project the moving images is accidentally knocked on its side, igniting several nearby oxygen bottles. Among the victims are some of the most prominent members of the French aristocracy, as well as such dignitaries as Duchess Sophie in Bayern, sister to Empress Elisabeth of Austria and the former fiancée of Ludwig II. The new 35 mm Normandin-Joly projector is held responsible for the conflagration, setting back the promotion of cinema in France by several years. Equally to blame, however, are the twisting alleyways and cramped cul-de-sacs of *le vieux Paris*. Swept away by Baron Haussmann and replaced by broad tree-lined boulevards and expansive gardens in a modernisation programme that saw 20,000 houses destroyed and over 40,000 new ones built during a twenty-year period of renovation between 1852 and 1872, medieval Paris is still a living memory at the time of the blaze. Reconstructed out of canvas and

26

48
62, 108

wood only three short years before the start of a century that will rapidly be dominated by the organisational power of cinema, it indicates the extent to which moving pictures and architecture have become nega-

18, 68

tive moulds of each other, particularly where disaster is involved. Both amplify its effects.

In cinematic and architectural terms, Tokyo's vast urban sprawl has long been regarded as defining the shape and nature of city life for the twenty-first century. The planners and engineers responsible for Tokyo's safety have described it as a 'disaster amplification mechanism': a term which, in a country where a seismic event occurs every five minutes, can also be applied to Godzilla himself. The unforeseen product of atomic tests taking place in the Pacific Ocean over sixty years ago, the blinding intensity of Godzilla's birth has ensured that he casts more than

52, 108

one shadow over the intervening decades. Whichever way you look at him, he remains the ultimate outcast. A giant prehistoric lizard trapped in a world overrun by tiny humans, a marvel of modern science that has no future, a powerful god who must be destroyed and, perhaps most shamefully of all, a man struggling to walk upright in a heavy rubber suit – in his time, Godzilla has been all of these things and more. 'Why has

69, 103

this monster appeared in Japan?' the noted palaeontologist Dr Yamane asks in the first ever Godzilla movie, released in 1954. Good question. Like all true monsters, Godzilla keeps himself removed from the real world: the damage he inflicts upon it is largely illusory. He continually destroys Tokyo only to see it rebuilt again each time as a larger and more detailed film set.

Movie scenery can never be anything other than temporary: the staging of the Bazar de la Charité against the burning backdrop of medieval

82, 116

Paris has seen to that. Fleeing crowds become the true stars of twentieth-century cinema. With the coming of Godzilla, for example, an entire generation learns to run backwards while pointing up at the sky and screaming. One such scenario has Michael Jackson, Elizabeth Taylor

75, 115

and Marlon Brando escaping New York in a hire car after the attacks on the World Trade Centre on 11 September 2001. With the grounding of all internal flights, they take it to be their only way of returning to the relative safety of Los Angeles and the West Coast. 'Brando allegedly annoyed his colleagues by stopping at every KFC and Burger King they passed along the highway,' *Vanity Fair* later reports. An increasing percentage of the planet's population is finding itself in transit without recognised papers or personal identity: a shifting human desert. The authorities in New York still have no clear idea of how many migrant workers perished during the events of 9/11 as there has never been any official documentation of their existence. Conversely, victims of the Bazar de la Charité fire in 1897 are among the first in history to be identified solely by their dental records. Individual characters, events and scenes are merely casual assemblages made up of gestures and perspectives – when it comes to our appreciation of architecture and cinema we choose not to notice how poorly these pieces fit together. The relationship between the human imagination and disaster has consequently never been a comfortable one.

• 87 •

You take a bus to a checkpoint about 20 km from the power plant, where you change to another bus which never leaves the contaminated zone. Among the 'major attractions' are the villages of Kopachi, which has 'partially encased' houses, and Pripiat, a ghost town left   25, 62 *exactly as it was when the residents were evacuated. The culmination of the tour is the cement- and lead-encased nuclear reactor, known as the Sarcophagus. The casing is so corroded now that you could actually put an arm through the lead and concrete at some points. After a final radiation check, visitors switch back to their original buses for the return to Kiev.*
    – Chernobyl Travel Guide

Godzilla's origins remain blurred, since he only exists to be interpreted. That is how any monster is finally called into being. A collision of meanings, Godzilla's presence is determined by how we make sense of him as a conflict between nature and science, the past and the future, destruction and security, the rural and the urban. Breathing radioactive fire onto burning buildings, he transforms cities into wastelands while helpless crowds stand and stare. A total of eleven reactors at power plants throughout Japan are automatically shut down in the wake of the 2011 Tohoku Earthquake: 'There is talk of an apocalypse,' Europe's Energy Commissioner Günther Oettinger declares, 'and I think the word is particularly well chosen. Practically everything is out of control. I cannot exclude the worst in the hours and days to come.' Large explosions and highly toxic leaks result in the forced evacuation of all residences within a twenty-kilometre radius of the Fukushima I Power Planet and a ten-kilometre radius of Fukushima II. Despite the risk of contamination, hundreds of survivors seek shelter from the earthquake's aftermath inside the nuclear power station at Onagawa.

Living with radiation becomes second nature in an age when energy and information are traded so freely together. Under such ideological pressure, atomic science is equated with alchemy. Both represent hermetically controlled operations that subtly and radically transform nature, thereby unleashing occult powers that remain hidden in all but their effect. The temporary storage of radioactive materials profoundly disturbs the natural terrain: it becomes reworked. The nuclear reactor is fixed in a landscape capable only of decay. Invisible yet exerting a powerful influence, its presence is best detected through the behaviour of those who have shaped their lives around it: the way fingers form themselves blindly around the handle of a cup in order to hold it steady. Within six months of the Fukushima meltdown over 18,000 personnel are reporting for work at the plant.

Nuclear reactors supply the archaeological sites of the future, but this has nothing to do with progress. All buildings contain within them the blueprints for their own destruction. 'I see Brasilia as I see Rome. Brasilia began with a final simplification of ruins,' recalls Clarice Lispector of her first encounter with Brazil's new federal capital. Ruins affirm the epic element in history even as they negate it. Like any enhancement to the landscape, they deceive the eye in order to please. Ruins therefore mark the spectacular return of nature, which is also the absence of

70, 113

77, 99

9

humanity – science is revealed in its semi-savage stage. As indicated in the Chernobyl Travel Guide, the atomic pile lies at the centre of a tourist attraction that must inevitably be deserted, so long as the prospect of disaster continues to hang over it. When an event has a half-life, everyone becomes a visitor. Radioactive contamination confers upon any scene a picturesque morality formerly associated with the earliest romantic landscapes, such as 'the garden that Mr Tyers had arranged for his leisure hours at Denbie, in which the paths were emblems of human life, now comfortable and flat, now rugged and steep, with here and there stones bearing moral inscriptions; and best of all, the Valley of the Shadow of Death, with coffins instead of columns, and skulls scattered picturesquely about'.

84, 111
47, 94

Ruins allow the landscape to speak in metaphor, interrupting the grand narrative of the past with portents of the present; and as with anything that clearly artificial, they exist only to entertain. Another statue in Jonathan Tyers's garden shows Truth trampling the mask of Illusion underfoot – perhaps a regretful reference to his role in the creation of the Spring Gardens at Vauxhall: one of the earliest expressions of Rococo horticultural design in England. With the introduction of landscape, nature is replaced by a system of values that renders it artificial: that is, subject to accusation, contradiction and finally annihilation. The gardens created at Versailles for Louis XIV of France, complete with a labyrinth featuring 'thirty-nine groups of hydraulic statuary representing the fables of Aesop' are described by one English aristocrat as 'a lasting monument of a taste the most depraved' despite being 'executed with boundless expense by the greatest artists of the age'. The artificial is therefore the preservation of that which may not exist, may not have existed and may never be permitted to exist. Preservation, however, alters the very thing it seeks to preserve – it is impossible to survive outside its influence. Entertaining and deceptive as it may be, the landscape's exterior remains wholly negative.

74, 114

An early proposal from Baron Haussmann for the reconstruction of Paris is the relocation of all existing graveyards to a purpose-built necropolis located on the rural outskirts of the city and linked to its centre by rail. The plan suggests a proper response to the monstrous observation made in a late-night cemetery by Lautréamont's Maldoror as Haussmann's plans approach their realisation: 'Gravedigger, it is fine to contemplate the ruins of cities, but finer still to contemplate the ruins of men!' In the horrific wake of the Tohoku Earthquake, a former bowling alley in a suburb bordering Sendai is used as a temporary morgue. More than 100 white coffins rest on the dividers between the twenty-five lanes at Airport Bowl. People walk up and down the empty lanes, looking into each coffin in search of loved ones – for each body identified, the Japanese Self Defence Force bring in at least another five; 'you just end up feeling guilty for living,' one visiting journalist remarks. As cinematic fantasy, a collision of projected meanings, Godzilla emerges simply as a landscape gone rampant: not natural at all but feral.

7, 119

• 88 •

Thus the science fiction film (like that of a very different contemporary genre, the Happening) is concerned with the aesthetics of destruction, with the peculiar beauties to be found in wreaking havoc, making a mess. And it is in the imagery of destruction that the core of a good science fiction film lies. Hence the disadvantages of the cheap film – in which the monster appears or the rocket lands in a small dull-looking town.

– Susan Sontag, 'The Imagination of Disaster'

18, 40

The makers of cheap science-fiction films, as Susan Sontag points out in her 1965 essay 'The Imagination of Disaster', have always favoured the desert; a landscape drained of meaning, caught between disuse and exhaustion, it offers the cheapest of cheap locations. Out here the aesthetics of destruction have been reduced to zero; Utopia and Oblivion are now one: the avant-garde happening and the science-fiction fantasy start to mirror each other. Amid scenes of crashing spaceships, alien invaders and rampaging giant insects, nature is finally

50, 79
42, 78

replaced by technology. The depiction of life on other planets only confirms this relationship, especially in the cheaper fantasies of science where the desert stands in both for the world we presume to know and for those we have yet to discover; none of them is capable of being comprehended except as a technological presence.

From Thomas More and Francis Bacon to Charles Fourier and Samuel Butler, Utopian literature has provided some of the earliest theoretical writings on the interdependence of humans and machines. Technology remains a key element in the establishing of each new Utopia first by correcting nature and then by displacing it. One convincing expression of this process is the landscape itself, which confers a moral – which is to say, utilitarian – interpretation upon the workings of nature. Everywhere you look, farmers and painters have been there before you. Utopian fantasies ultimately satirise this condition, no matter what the poets and philosophers might claim, and technology plays an important part in an understanding of such satirical treatment. In the words of WikiLeaks spokesperson Jacob Appelbaum: 'I think there are a lot of us who don't think the world is as we would like to see it. And anyone who is not a utopianist is a schmuck.' 51, 101

Nature has always been valueless except as a series of absolute forms from which everything else is copied. As such it remains one of the earliest models for analogue communication in existence. On 27 February 1781 the Eidophusikon, meaning 'the imitation of nature' in the language of all dead media, opens in Leicester Square for the first time. 75, 96 Devised by the landscape painter and theatrical set designer Philippe Jacques de Loutherbourg, it is presented to the public as offering nothing less than 'MOVING PICTURES, representing PHENOMENA of NATURE'. Within a landscape frame whose dimensions are little more than seven feet by four, a series of animated scenes is staged, accompanied by sound effects, music and the recitation of poetry. Scrims, screens, mirrors, painted backdrops and coloured lights are deployed to recreate a storm at sea, rolling moorland vistas – or most spectacularly of all, Satan summoning his army of demons by the burning lake of Hell, as described in John Milton's *Paradise Lost*. Members of the Eidophusikon's audience have been known to faint before such carefully contrived illusions.

Fainting under such circumstances says less about the illusion itself than about what it seeks to convey. How else to explain such extreme behaviour? The Eidophusikon's audiences must have found – or lost – themselves in a theatre before. For an age in which landscape is increasingly to be regarded as a sublime spectacle, their response is the correct one: only the crudest and most prosaic of individuals could stand before a beautiful sunset, seascape or pastoral scene without fainting. It is the sign of a cultivated sensibility. Loutherbourg's influence is felt in the set designs at David Garrick's Drury Lane Theatre as much as in the heightened colour and dramatic lighting effects of his friend Thomas Gainsborough's later landscapes. The Eidophusikon's fainting audiences draw attention to themselves because of a shift in convention: 'the imitation of nature' seems to require a corresponding imitation of response.

· 89 ·

It is conventional to call 'monster' any blending of dissonant elements; the Centaur and the Chimera are defined this way to people who do not know them. I call 'monster' every original inexhaustible beauty.
        – Alfred Jarry

As a close reading of 'The Imagination of Disaster' will reveal, cheap science-fiction films exploit similar shifts in convention in order to reassure their audiences: according to Sontag's argument, technology will not permit the crashed spaceships, giant bugs or alien invaders out in the desert to cause us any lasting harm because it is technology that placed them there in the first place. Satan and his demon hordes remain trapped beside the fiery waters of Hell by the very stage machinery that conjures them up. To the extent that they can be interpreted, Sontag seems to imply, such illusions have no effect – only meaning.

There is no cause for alarm – only a numbing sense of the inevitable. 'The films reflect world-wide anxieties,' she writes in and of an age terrified of nuclear oblivion, 'and they serve to allay them. They inculcate a strange apathy concerning the processes of radiation, contamination, and destruction which I for one find haunting and depressing.' The year before this remark first appears in print, building is completed on the Economist Plaza in the centre of London. The work of Peter and Alison Smithson, the basis for this architectural project is an elevated precinct upon which towers of different heights but similar design are carefully positioned. Each of them clad in the same roach limestone bearing the pockmarks left behind by fossilised creatures, they stand at the centre of a new economic world that takes the form of a vast desert of concrete and sheet glass – as if the future has already happened.

'This change of scale between the two buildings,' declares *Architectural Design*, regarding an early visit to the Economist Plaza, 'has resulted in a giant *trompe-d'oeil* with which one is only to experience further perceptual difficulty as one enters the centre of the plaza. In the centre of the plaza the "photographic" reduction in scale of the residential block *viz-a-viz* the main tower has the optical effect of "zooming" this block away from the observer, with a consequent dramatic enlargement in the apparent space of the plaza. This perceptual sleight of hand is brilliant but not in the last instance felicitous, for the observer does not remain rooted in the centre and moving on he quickly discovers the deception. Once the residual illusion of the residential building has been exposed, the whole assembly is open to "theatrical" interpretation, and this interpretation does not help in sustaining belief in the monumentality of the major office tower.' In other words, it is easy to forget that our understanding of what constitutes an illusion is usually expressed in architectural terms but understood in cinematic ones.

Illusions require depth in order to be credible; behind its seven-by-four proscenium frame the Eidophusikon's stage stretches back a full eight feet. The resultant compression and falsification of perspective brought about by the shift from two dimensions to three reassure the eye even as they deceive it: a device further exploited in the 'zooming' away, 'photographic' reduction in scale and perceptual sleight of hand on display at the Economist Plaza. The Eidophusikon anticipates later scenic illusions, such as Louis Daguerre's dioramas of the 1820s, and the popular historical panoramas of the late nineteenth century – just as the Economist Plaza reflects a growing preoccupation with the anamorphic film projection systems, such as Cinemascope and Cinerama, and anticipates a recurring fascination with movies shot and viewed in 3-D. We now take the 'original inexhaustible beauty' of illusions for granted, ignoring its cultural codes only because they continue to work too well upon us. Isn't a social history of the cinematic effect therefore long overdue? There is, for example, the story about a security guard in a Tokyo department store attempting to arrest two suspected terrorists who have been overheard by shoppers animatedly discussing the detailed destruction of the Japanese capital. The dangerous subversives turn out to be veteran film director Ishirō Honda swapping ideas with special effects expert Eiji Tsuburaya for a scene in what will become the first Godzilla movie; and the city they are planning to destroy is being built in 1/24 scale out of plaster and balsa wood on a Toho studio soundstage.

22

75, 99

• 90 •

78, 102

'No, you were born in Tokyorama – the Land of the Rising Sun.'
        – Lemmy Caution, Alphaville

The Athenians would erect altars dedicated 'to an unknown god'. Illusions are customarily the result of not being in full possession of the facts: they owe their existence to whatever we choose to leave hidden behind the curtain. It is therefore wise to consider those gods whom you may have overlooked or never see, as they too will demand their sacrifice. The Minotaur concealed by his labyrinth and Moloch, lurking deep within the giant subterranean machine rooms of Metropolis, lie in wait to devour their human victims. Those who wish to remain unseen can have no stable form: only scenes of death and disaster will reveal them. At the start of his first film Godzilla is presented to his audience as

49, 120
53, 98

a series of destructive forces: a blinding atomic flash, a wrathful deity, a violent nocturnal storm. The giant monster takes shape and gradually becomes visible as he makes his relentless way towards Tokyo. Outlined against the flames of a hundred burning buildings amidst the panoramic display of the city's ruins, Godzilla's presence is finally at its strongest.

Godzilla destroys at night. The terror he inflicts is at its starkest and most dreamlike when viewed in black and white. 'Godzilla is approaching Tokyo,' a radio announcer politely informs his frightened listeners as a monstrous silent form emerges from the dark waters of the bay. Modern media, in the form of masts and transmitters, pylons and cables, are quite useless in stopping him. Like the television news camera that starts out following Godzilla's rampage from the top of the Tokyo Tower only to end up relaying its own fall to the ground below instead, nothing can capture his true nature quite like monochrome film. Technicolor inevitably proves to be his undoing; suddenly it reveals that Godzilla's skin is not green but a dark grey, like the colour of earth that has been hit by a powerful explosion. Tohoscope widescreen cinematography magnifies each disaster he creates into intricate spectacle.

Greater popularity also means that Godzilla is obliged to emerge into the light of day. By the mid-1960s, he has become such a star that the buildings he topples along the Tokyo skyline tend to be empty and no one seems to get hurt anymore. People read in their newspapers about the terrible destruction that is taking place at that moment in another part of town. Television and radio rarely fail to respond to the monster's 80, 100 presence now: they follow his progress from across the city. His gestures have now been caught so completely on camera that the same scenes of running crowds are repeated from movie to movie. Under an eternal blue sky, Godzilla keeps knocking Tokyo down, and it is constantly being rebuilt again. He becomes so closely associated with the city's progress that it now appears as if its expansion, both upwards and outwards, is based entirely upon his efforts. Eventually, Godzilla finds himself the protector of a world that has no place for him except as its would-be destroyer.

Like any unknown god, he is best understood in terms of the very thing he seeks to destroy. At the end of his first film, as Godzilla disappears back into the ocean, his presence grows indistinct once again. He is, however, also followed into the depths of Tokyo Bay by Dr Serizawa, costumed in a deep-sea diver's grey rubber suit and brass helmet for the encounter. An embittered and reclusive scientist, Serizawa has grudgingly agreed to use his latest invention to kill the monster. Referred to as the 'oxygen destroyer', the device will transform the surrounding seawater into a violently churning cauldron which will eventually claim the lives of both Godzilla and Serizawa. Choosing to perish rather than reveal the secret of the oxygen destroyer, Serizawa cuts his own airline, severing all links with the surface; man and monster die together

62, 111   amidst foaming columns of bubbles and shadows. Only in scenes of death and disaster will the unseen finally reveal itself. Born out of a thermonuclear explosion then dissolving into a lifeless sea, the awful truth is that Godzilla represents human evolution run backwards. He is, in other words, precisely what he appears to be on the screen: a man in a grey rubber suit.

· 91 ·

As a factory – the factory is right. As life – the factory is a flop.
  – Viktor Shklovsky, Third Factory

According to a persistent myth, special effects director Eiji Tsuburaya resorts to a man in a rubber suit to play Godzilla solely because he lacks the necessary expertise to reproduce the stop-motion technique de-

7, 62   veloped in Hollywood by Willis O'Brien for the making of *King Kong*. A close examination of the first Godzilla film reveals that a complex amalgam of techniques has been used to bring the monster to life, ranging from highly detailed miniatures and matte shots to wirework, marionettes and glove puppets. Stop-motion does in fact feature in *Gojira*, but only to convey impressions of violent or rapid motion, such as the flicking of the monster's tail or a high-speed automobile collision. The decision to restrict its use is entirely a financial one: stop-motion being so costly and time-consuming that even the major Hollywood studios tend to avoid it when they can. What was heralded at the start of the twentieth century as the Seventh Art has quickly become a Seventh Factory. There has never been a film industry: only film   75, 98 as industry. Cinema lends itself so readily to this process. Each Godzilla movie has only ever been a machine designed to produce a specific version of the same monster using an established set of methods.

Godzilla dies his own death at the end of each film only to be resurrect-   79, 103 ed, just like Tokyorama, in the next. He never comes back as the same creature, however. Although Godzilla owes his birth to atomic science, and to the nightmares engendered at Hiroshima and Nagasaki, he remains the product of something equally powerful in its own way: the studio system. During the 1960s Godzilla becomes the ultimate 'salaryman': a corporate worker and team player who works long and hard on behalf of his company. Under Toho's direction, Godzilla's facial features and character have been reconstructed from movie to movie: rubber perishes and must be replaced, after all. The post-war salaryman is the subject of hundreds of movie comedies gently satirising the life of the modern office worker; together and in their own separate ways, Godzilla and the salaryman have helped to make Tokyo what it is in the second half of the twentieth century. Thus *King Kong vs. Godzilla*, released in 1962 by Toho as a salaryman comedy, does not simply play out a conflict between rival monsters but different animation techniques. Most evidently a man in a gorilla suit, King Kong finds himself in Godzilla's world now.

Which leads us to the inevitable question: can Godzilla ever be digital? Is it possible for him to exist as a computer-generated image? In an age when cultural identities are becoming increasingly globalised and products more carefully branded, Godzilla reveals the extent to which illusions still depend upon specific cultural codes in order to be effective. With the introduction of digital VFX and 3-D projection systems, these conventions are flipped over and forcibly imposed upon the audience. We are not by nature digital creatures – on the surface at

least. How did the pleasure at being tricked into believing – our willingness to accept optical illusions – turn into the enforced consumption of spectacle? Technology finally comes to represent the economic
and strategic exploitation of thought and action. In the end we run the
risk of consuming ourselves. We have no choice but to surrender to an
unwilling suspension of belief; even lens flare and other optical side-
effects can be programmed into the digital panoramas of marauding
monsters, giant robots and collapsing skyscrapers currently playing at
our local Cineplex. In the Information Age the dream factory is just a
factory: the crawl of names credited at the end of each production confirms as much. Thanks to 'motion capture' technology, we may still find
ourselves looking at the very thing that Godzilla was always derided for
being in the West: a man in a grey rubber suit.

24

• 92 •

Bigger than the screen, destroyer of entire buildings, Godzilla does his
worst damage from within the cinema's darkened interior. The district
of Tokyo that Godzilla is shown attacking in his first film is the one immediately outside the picture house in which it premieres. The audience
watches as he crushes its roof underfoot. They are, however, already
prepared for this moment. In Japan a radio adaptation of *Gojira* runs
from July to September in 1954, prior to the film's eventual release in
November. Once the lights go down in the cinema, the assault on the
audience's ears continues. The heavy pounding of Godzilla's footsteps
is heard over the opening credits, followed by his angry roar. To help
convey the impression that Godzilla is about to smash his way into the
auditorium, the monster's roars are blended in with the film's main title
theme, an urgent, relentless march composed by Akira Ifukube. People
consequently hear the monster coming long before they see him. He
will not appear again until another twenty minutes into the movie. A scientist investigating Odo Island switches on a Geiger counter; it crackles
and Godzilla suddenly appears over the brow of a mountain. And in the
next moment he is gone again.

Sound is a promiscuous entity, giving the active illusion of being able to cross boundaries and traverse space. Music's ability to define itself architecturally – and, by extension, geographically as well – is one clear aspect of this phenomenon. The promiscuity of sound captures moments of transition and disappearance with the greatest agility. The clicking of the Geiger counter needle, registering radioactive contamination in the air, directly connects the various elemental forms that Godzilla has assumed during the early stages of *Gojira* with the film's soundtrack of collapsing buildings and the eventual emergence of new city towers. We hear what is coming: we only see what is before us. Depicting the world of the ancient gods consumed by fire and flood, Wagner's *Götterdämmerung* offers destruction as music drama, ending in the final immolation of Valhalla. The epic demise of the gods is accompanied by an epic theme of orgiastic renunciation, longing and loss. Soaring above the flames, this final leitmotif is an echo not only of all that has gone before it in the Ring Cycle but also a prelude to the music dramas that are yet to come.

3, 76
48, 96

The leitmotif sweeps everything before it in a thematic series of echoes which, through strict repetition, fixes the listener's conscious attention not just on what has happened but what is about to happen. As a form of narrative reverberation it has an immediate counterpart in Wagner's dramatisation of acoustic space. Instruments and voices are isolated within the opera house's static dimensions and thereby given the illusion of motion: hunting horns call and respond to each other from either side of the proscenium arch; pilgrims and courtiers process across the stage in organised columns, and lone voices call from beyond its depths. Just as the Smithsons' design for the Economist Plaza organises space cinematically through a precisely placed sequence of visual echoes, so Wagner's music dramas remain predominantly spatial, no matter what appears to be happening in front of the audience's eyes. It is therefore probable that even after film has ceased to maintain itself as an art form it will continue to survive as music. In this way, its influence will be felt long after the unstoppable torrents of today's Digital Regime have swept cinema away, leaving only a set of downloadable audio-visual components in its wake. Digital sound recording and projection

85, 106

techniques have meant that movies are, in fact, growing louder; thanks to transducers and compression systems such as Dolby and THX, Wagner's acoustic illusions have been extended to embrace explosions and impacts that seem to shake the entire auditorium without apparently harming anyone.

· 93 ·

The necessity of realistic scenarios and backstory in military simulations led designers to build databases of historical, geographic and physical data, reconsider the role of synthetic agents in their simulations and consult with games design and entertainment talents for the latest word on narrative and performance. Even when this has not been the intention of their designers and sponsors, military simulations have been deeply embedded in commercial forms of entertainment, for example, by providing content and technology deployed in computer and video games.

> *– Tim Lenoir and Henry Lowood, 'Theaters of War'*

Digitally generated armies that can learn to fight, kill, receive wounds and fall down convincingly swarm across the movie screen in virtual hordes, acting out panoramic scenes of crowd violence in which not a single human being has directly participated – their brief onscreen lives are barely noticed. At the same time 'kinetic military action' has become a recurring government euphemism for actual combat, as if the actual transfer of data and energy causes the greater harm. Tim Lenoir of Stanford University has argued for the emergence of a 'military-entertainment complex' of effective combat simulation and interactive gaming design. Ever since the run-up to Desert Storm, however, it would perhaps be more accurate to suggest that commercial forms of entertainment have been deeply embedded in military simulations than the other way around.

Seconded from the US Air Force to DARPA, Colonel Jack Thorpe

13, 64

57, 104

77, 102

reconnects Carl von Clausewitz to the military mainstream via SIMNET, the first computerised simulator network, initially used in preparations for Desert Storm. What emerges is not the faithful copying of an existing piece of weapons technology – what is sometimes called an 'airplane on a stick', capable of simulating the behaviour of a specific device – but rather a digital 'design for effect' focussed upon the crews and units of personnel involved in a full-scale military operation. They now find themselves within a simulated landscape that is perceivable as a series of interactions, timetabled events and reaction times. 'The simulators themselves,' Lenoir observes, 'presented synthetic environments – virtual worlds – by utilizing advances in computer graphics and virtual reality research. With the rapid development of Distributed Interactive Simulations technology during the 1990s, content and compelling story development became increasingly important.'

64, 117

As our armed forces increasingly operate online – the presence of the Enemy inside the network having long since become the normal state of affairs – a new convergence between Hollywood VFX experts, videogame designers and military strategists has been brought about. Everything from specific mission training and the test deployment of expensive new weapons systems to the treatment of Post Traumatic Stress Disorder is taking place within simulated environments. 'It's something that we take for granted today,' Thorpe explains, 'because the SIMNET technology was the existence proof for the beginning of network gaming in the late eighties and early nineties. As microprocessors became inexpensive enough that you and I could actually own one, graphics engines in the microprocessors allowed us to generate a pretty interesting visual world, and networks actually started to proliferate so that I could connect to you. That was the beginning of network gaming, which led to the massively populated persistent worlds, these massive online games that have hundreds or thousands of people playing, which has gone back now and impacted military training.'

The videogame's mix of disposable excess and high-end playability facilitates the assimilation of terror with normality and the growing perception of destruction as the ultimate playpen for the human imagination. The Joint Fires and Effects Trainer System (JFETS) installed at Fort Sill in Oklahoma uses gigantic wall-mounted HD screens, a super-woofer sound system and interactive processing to recreate Iraqi combat scenarios partly as an arcade game and partly as an IMAX spectacle for Veterans of Future Wars who have grown up with precisely these kinds of media entertainment. The result is a 'panorama on a stick' in which the Enemy is presented as a series of projected images and digital sound recordings. In this way the battle has been logistically separated from the battlefield; and communications technology creates yet another new landscape. The JFETS simulator is a product of the Institute for Creative Technologies, an army-funded research and development group established at the University of Southern California under the directorship of Richard Lindheim, a former Paramount TV executive.

3, 65

• 94 •

What has happened is – now you all have to turn your brains around – the greatest work of art there has ever been. That minds could achieve something in one act, which we in music cannot even dream of, that people rehearse like crazy for ten years, totally fanatically for one concert, and then die. This is the greatest possible work of art in the entire cosmos. Imagine what happened there. There are people who are so concentrated on one performance, and then 5000 people are chased into the Afterlife, in one moment. This I could not do. Compared to this, we are nothing as composers...
  – Karlheinz Stockhausen, 18 September 2001

11 September 2001: American Airlines Flight 11 and United Airlines Flight 175 smash into the Twin Towers of New York's World Trade Center, while American Airlines Flight 77 strikes the Pentagon and United Airlines Flight 93 crashes into a field in Pennsylvania. Karlheinz Stockhausen's remarks, made at a Hamburg press conference barely a week

12, 80
58

later, are intended to be allegorical; the composer maintains that he 'used the designation "work of art" to mean the work of destruction personified in Lucifer' as represented in *Licht*, his cycle of seven operas. 'In the context of my other comments this was unequivocal,' he adds. Modern playback systems bring bodies and sounds together at greater velocities; but what we see is still determined to a large extent by what we hear. Where Wagner's dramatisation of acoustic space creates the illusion of movement by isolating sounds in space, Stockhausen introduces an even greater dynamic by shifting it backwards and forwards between the left and right channels of a stereophonic speaker system, thanks to a Doppler effect he achieves at the WDR studios in Cologne by slowing down and speeding up individual tones on magnetic tape. We only hear what is coming.

'Yeah, of course it is always the case,' Stockhausen observes in 1999 with regard to the demolition of one World Expo site so that another can take its place somewhere. 'I read yesterday that some asteroids almost crashed into this planet a few years ago. There was a danger that this could happen, and the whole planet would be smashed to pieces. So we would have to start somewhere else.' Universal expositions offer a temporary location for the world of tomorrow before becoming its final resting place; carefree tourists get to wander in a future that is both briefly glimpsed and carefully controlled. Are monsters the only ones to break our web of purposes? And do we recognise them solely by the ruins they leave in their wake? Photographed together at the 1958 Brussels World's Fair while standing behind a painted canvas screen, John Cage and Karlheinz Stockhausen both appear to be flying in the same replica 'airplane on a stick'.

6, 87

20, 76

Too soon? Geologists suggest that we are entering a new era, the 'Anthropocene', in which the human impact upon the planet's development is forming its own stratified layer as distinct as that of the Jurassic or Cambrian periods. The 'man-caused disasters' designated as acts of terrorism suggest the extent to which massed humanity has become the medium through which 'the work of destruction' communicates itself. Meanwhile the Japanese government unveils its plans for a new city, codenamed 'IRTBBC' or 'Integrated Resort, Tourism, Business and Backup City', to stand in for Tokyo in the event of the nation's capital being destroyed in an earthquake. 'IRTBBC will incorporate all the vital functions of government,' according to one report, 'with duplicate facilities for parliament, ministries but also include office complexes, resort facilities, casinos, parks and the tallest tower in the world at 652 metres.' A possible location lies 300 miles west of Tokyo on land currently occupied by Itami Airport.

· 95 ·

As the gods become completely hidden from view by the flames, the curtain falls.
    – Richard Wagner, *final stage direction*, Götterdämmerung

Godzilla is dead, and we have killed him. Analogue special effects brought a certain clarity to cinematic illusion, isolating both monster and ruins against the same background. In a sequence from *Gojira* depicting a jet-fighter attack on Godzilla, tiny model planes are clearly seen hitting a backdrop set up at the edge of the soundstage to represent the sky and dropping in flames onto the studio floor. Digital illusion, however, is a devastating assault upon the senses: the most spectacular VFX create the biggest onscreen disasters as each new simulated world comes crashing down around our ears. Rapid edits keep audiences permanently disorientated; endlessly scrolling computer-generated action sequences replicate videogame interactivity. Unable to interpret what is now before them, we can no longer believe our own eyes. Digitally rendered distortion and camera-shake, blurred movement and distorted sound stand in for 'reality' at its rawest. At this point, it is safe to assume that the audience has completely taken leave of its senses and the cinematic experience which had first shaped them no longer exists.

In the Digital Regime clarity renews itself not on the cinema screen but in the perceptions of the audience, particularly as they apply to the world outside. With the introduction of the 'citizen reporter' posting low-res images of bomb attacks, riots and disasters online, the digital fragmentation that accompanies recorded explosions has become the new cipher for actuality. With immediacy granted the authority of being an 'event' in itself, visual confusion is the measure of experiential reality. Nature is perceived as noise: which means that we seek to correct it, render the wilderness less 'wild' and more interesting. This is happening quicker than the eye can capture. The line between abstraction and representation is rapidly becoming indistinct. Images exist that can never be completely resolved. The more comprehensive the digital illusion, the more cinematic experience itself becomes blurred, distorted, smeared and fragmented. It can only exist at the outer edge of the Digital Regime, as if the binary weave of ones and zeros were unravelling or being picked apart, rendering the organisational power of cinema suddenly incomprehensible and its effects upon our senses indiscernible.

The digital image has all the picturesque appeal of a landscape that has been artfully strewn with ruins. Its final simplification expresses itself as a 'will to imperfection' which equates nature with violence and cinematic experience with disaster. A whole new genre of movies, supposedly edited together from jerky 'found footage', has subsequently come into existence. The studio panic that opens George Romero's *Dawn of the Dead* gives way in *Diary of the Dead* to a group of film students out on location in the woods making a low-budget horror film; upon hearing media reports of rioting, mass-murder and cannibalism, they turn their video cameras upon themselves, documenting a 'real' zombie attack. The raw scenes of violence, chaos and death that they capture are later cut together to form the secret narrative – 'the true story' – presented to the audience. As 'the ruins of men' that continue to walk among men, zombies are creatures without a god and therefore created in nobody's image. No wonder the zombie has adapted itself so neatly to existence online, taking on a mindless viral presence of its own.

68

The final simplification of ruins will always be digital. 'What you mean is,' Lemmy Caution curtly observes, 'it's not Alphaville you're prattling on about… it's Zeroville. And what on earth are we going to see anyway?' What else? *Alphaville* is the film that Alpha 60 finally makes of its own destruction: in a networked city people either behave in an orderly pro-grammed manner or flail about in uncontrollable panic. From the very start the events of 9/11 are compared in the mainstream media to scenes from a disaster movie: as if the world itself had changed and we had only just noticed. Disasters, as Maurice Blanchot has noted, ruin everything, all the while leaving everything intact – which is another way of saying that a disaster will always be local and our understanding of it equally so. The instantaneous global dispersal of information assures us of this. The network forms itself onscreen into a series of computer-generated spe-cial effects: gigantic structures, organisms and machines are constantly falling apart then reassembling themselves. Cities are destroyed, which is to say digitally transformed: their 'web of purposes' exposed as the circuitry of information running through them.

3, 79   From the irradiated wastes of Chernobyl to China's deserted theme parks and the yellowing remains of America's abandoned suburbs, on-line images of urban ruins, abandoned amusements and polluted sites offer access to a new order of shapes and spaces. Located in Outer Mongolia, Dinosaurs Fairyland is an empty tourist attraction featuring full-scale fake dinosaurs smashing down walls or hatching from eggs amidst skeletal remains both real and faked; the park's name is written
7, 62   in giant letters copied from the Hollywood sign, spelled out in both Eng-lish and Mandarin, and military vehicles run along its dusty pathways. In the Gobi desert the abandoned city of Ordos is a planned community for one million residents where nobody can afford to live: a new local museum has been added to serve a population that still refuses to exist. The crenellated walls and crumbling fairy-tale towers of Wonderland, the Chinese version of Disneyland, preside over acres of depopulat-ed farmland, while the Enchanted Forest amusement park of Ellicott City, Maryland, the first ever to open in the United States, rots away be-hind locked gates. With its decaying malls and theatres, the legendary zombie architecture of Detroit, where a third of all homes are vacant,

constitutes all that remains of a disaster movie once the cameras have stopped turning.

In major cities throughout the world, computer-controlled firework displays mark the start of each New Year with carefully organised scenes of harmless devastation: a grand reconstruction of the Twilight of the Gods in which offices, stores, apartment blocks and landmarks are hidden by the synchronised projections of glowing smoke and coloured flame. Who would have guessed that the Apocalypse might take so long? *'Cornegidouille!'* writes Alfred Jarry in his guise as the monstrous Ubu from deep inside Haussmann's Paris. 'We would never have demolished everything if we hadn't meant to destroy the ruins as well. And I see no other way of doing that than by erecting some fine new edifices.' Godzilla and his crew of fellow monsters are an early expression of the sensory energy that has been tearing down the walls of the physical city, preparing us for its new existence as a series of networked experiences. 'Look the necessity full in the face, and understand it on its own terms,' John Ruskin admonishes a full half century before Jarry's words. 'It is a necessity for destruction. Accept it as such, pull the building down, throw its stones into neglected corners, make ballast of them or mortar, if you will; but do it honestly, and do not set up a Lie in their place.' In December 1996, Dr Yasuyuki Shirota of Hirosaki University announces plans to regenerate the moa, a giant flightless bird extinct for over a century, by using DNA extracts and live chicken embryos. Ultimately, Dr Shirota hopes to recreate a living dinosaur. His inspiration, it turns out, is watching Godzilla movies as a child. But why stop there? Why stop anywhere? God does not exist – yet. The curtain has only just started to fall.

36

Experimental Cognition: How to Pass the Turing Test

'For a century we have been preparing for an absolutely fundamental convulsion... Who will set up the image of man when men...have all fallen to the level of the animals or even of automata?'

Friedrich Nietzsche
Untimely Meditations III

• 96 •

Hereupon, the negro, grumbling out an apology, went up to his master, opened his mouth with the knowing air of a horse-jockey, and adjusted therein a somewhat singular-looking machine, in a very dexterous manner, that I could not altogether comprehend. The alteration, however, in the entire expression of the General's countenance was instantaneous and surprising. When he again spoke, his voice had resumed all that rich melody and strength which I had noticed upon our original introduction.

– Edgar Allan Poe, 'The Man Who Was Used Up'

It is only after the insertion of an artificial palette into his mouth by a trusted manservant, and the corresponding change in his voice, that the final revelation about the central character in Poe's tale can be made. 'It was evident. It was a clear case,' gasps the astonished narrator. 'Brevet Brigadier General John A. B. C. Smith was the man – was *the man that was used up.*' The deliberate repetition of 'was the man' marks the emphatic end to this macabre literary squib. Horribly mutilated during the 'Bugaboo war', the General's handsome physique and commanding presence have been restored to him only through 'the rapid march of mechanical invention' in the form of manmade limbs, eyes and torso. Any and all attempts to answer the narrator's questions about the distinguished officer's true condition have so far been interrupted by different possible interpretations of the word 'man',

which serve to connect together a series of seemingly unrelated statements. With a name so common that it borders on anonymity, saved
only by the alphabetically-organised eruption of middle initials at its
centre, John A B C Smith is more than just 'the man': he is McLuhan's
Typographic Man played out and exhausted as the most logical conclusion of a printed text. 'Used up' and capable of action solely through the
introduction of the most extreme prosthetics, he is nearly all extension
and very little else. Stripped of his godlike and numb exterior, all that
remains of the masterful John Smith is 'a large and exceedingly odd
looking bundle of something' that lies helplessly on the floor addressing
his servant in a voice 'between a squeak and a whistle' – and all before
the nineteenth century has even reached its midpoint.

64, 112
75, 120

With human identity rendered unfixed and formless through the introduction of machine prosthetics, beginnings and endings are easily
confused. According to a stage direction late in Karel Čapek's play *Rossum's Universal Robots* the Twilight of Humanity begins more or less as
Götterdämmerung ends. 'A thunderous tramping of feet is heard,' the
last line of the final Act proclaims, 'as the unseen Robots march while the
curtain falls.' First performed in Prague in 1921, Čapek's comedy of science is usually known by the typographically symmetrical arrangement
of the initials in its title: *R U R* – a contraction that Brevet Brigadier General John A B C Smith would have fully appreciated. With a past that has
been literally hacked to pieces, he can only look forward to the rapid
march of mechanical inventions yet to come. The unseen tramping of
robot feet is the last thing any of us will hear, however. Having cut down
the entire human race in a single violent uprising, the robot becomes
the prosthetic extension of humanity itself, even giving birth to its own
kind. This mechanical preservation of life is also a psychological denial
of death: a form of godlike repression that helps to distract attention
from that shapeless bundle of human matter at our feet calling for its
servants in a voice caught somewhere between a whistle and a squeak.

53

48, 92

Incapable of either walking or talking, the Robot Baby remains one
of the most complex expressions of 'used up' humanity and its
relationship to the mechanical prosthetic. Various incarnations have

already crawled about, staring at the world through artificial eyes. Pointing at colours and shapes, one Robot Baby learns to separate recognisable human words from the meaningless stream of phonetic babble it has been programmed to produce. More interesting still are the language games performed by infant robots in which they agree upon their own made-up names for the colours and shapes they point out to each other. Capable of learning through trial and error, the Robot Baby can only exist as a set of limits, no matter what its potential may be. It will always be rejected in the end. 'My Real Baby', the commercial version of a Robot Baby prototype, is no longer on the market, despite confident predictions that it will 'own Christmas 2000'. Sensors beneath its lifelike skin respond to a child's touch, causing it to giggle as if tickled and ask for a bottle when 'playtime' is over. A little over a hundred years previously, the Edison Talking Doll was pulled off the market in 1890. With the first cylinder phonograph intended for home use mounted inside her chest, the metal and porcelain figure was able, at the turn of a handle, to recite selected nursery rhymes or even wish its owner a 'Happy Christmas'. The doll proved to be an utter failure: hardly any sold and most of those were subsequently returned by disturbed customers. Even Edison had to admit that 'the voices of the little monsters were exceedingly unpleasant to hear'.

11, 88

65, 119

· 97 ·

Her step is peculiarly measured: all of her movements seem to stem from some kind of clockwork. Her playing and her singing are unpleasantly perfect, being as lifeless as a music box: it is the same with her dancing. We found Olympia to be rather weird, and we wanted to have nothing to do with her. She seems to us to be playing the part of a human being, and it's as if there really were something hidden behind all of this.

– E T A Hoffmann, 'The Sandman'

How shall we educate the Robot Baby? Will it grow to unsettling perfection solely through the manipulation of trial and error or, to examine the problem from an opposing perspective, out of the all-too human desire to be a machine? And how does this desire manifest itself as a historical process? Does the fantasy or the technology come first? In Hoffmann's 'The Sandman', a tale first published in 1816, Nathaniel is driven to madness and eventual suicide by the discovery that Olympia, the perfect object of his adoration, is a life-size clockwork automaton. Created by the sinister German craftsman Coppelius, Olympia demonstrates her clear superiority over Edison's Talking Doll by not speaking at all. The most she will ever murmur is an ambiguous 'Ah, ah!' This, however, is more than enough to encourage a young romantic like Nathaniel to believe that she completely understands everything he breathlessly confides in her. Endowed with great personal beauty and a name that connects her to the home of the ancient gods, it is Olympia's 'unpleasantly perfect' dancing, singing and playing and not her muted conversational skills that provoke the most unease – which in this case is nothing but the rationalised expression of an unacknowledged desire. Twenty 75   years later, Charles Babbage ruefully notes that it is a dancing female automaton, 'her eyes full of imagination and irresistible', which excites the interest of his party guests rather than the Analytic Engine, which sits largely ignored in an adjacent room.

Cognition begins with an awareness of one's own mistakes, consciousness itself being derived from the slow process of trial and error. Experimental Cognition is therefore an expression of how that consciousness interacts with technology and is in turn shaped by it. 'He dreamed of a waxworks shop,' Nietzsche writes in *Untimely Meditations* of a friend's experience, 'the classic authors stood there delicately imitated in wax and gems. They moved their arms and eyes and a screw inside them squeaked as they did so.' The ability to coexist contentedly with smart machines has little to do with rationality – that is to say, with perceived 83, 118   notions of sanity or intelligence. What is required instead is the appro-2, 51   priate delusion: what Walter Benjamin identified in an earlier age with 18, 51   the 'life-sustaining lie'. This defining characteristic of the Jugendstil line is also the most beguiling feature of the nineteenth-century automaton in that each 'represents an advance, insofar as the bourgeoisie gains access to the technological bases of its control over nature; a regression, insofar as it loses the power of looking the everyday in the face.' The formless and messy end to all human existence is replaced by the glamour of mechanical perfection. 'The bourgeoisie senses that its days are numbered; all the more it wishes to stay young,' Benjamin concludes. 'Thus it deludes itself with the prospect of a longer life or, at least, a death in beauty.' Only by misunderstanding technology do we permit ourselves to live with it on a daily basis. Experimental Cognition is consequently nothing other than the science of mistakes.

• 98 •

The robot has become the carrier signal for everything we fail to recognise as human – and, as such, it can never be what we want it to be. 'Those who think to master the industry are themselves mastered by it,' Čapek tells the *Saturday Review* in 1923, two years after *R U R* is first staged. 'Robots must be produced although they are a war industry, or rather *because* they are a war industry. The product of the human brain has escaped the control of human hands. This is the comedy of science.' Named after the Czech word for 'forced labour', the robot begins as a purely literary invention whose presence inevitably dominates Čapek's most famous play. Off the printed page, however, the robot is destined to remain an experiment in progress: a process without a recognisable end. All robots are consequently typographic in nature, caught between design and performance, text and machine.

Little more than a sophisticated printing press on legs, the robot inevitably functions best when subject to the established principles of mass production: standardisation and repetition. With the introduction of Frederick Winslow Taylor's *Principles of Scientific Management* during the early part of the twentieth century, the practical implications underlying Čapek's comedy are already apparent. The standardisation of production leads to the standardisation of parts and the assembling of

66, 113   more complex mechanisms. The assembly line emerges from the coercive harnessing of controlled spaces and routines designed to integrate metal and flesh. This will eventually permit Henry Ford to record in 1923 that, of the 7,882 actions required to assemble a Model T, '670 could be fulfilled by legless men, 2,637 by one-legged men, two by armless men, 715 by one-armed men and ten by blind men.' Experimental Cognition has nothing to do with the adaptation of man to machine or machine to man, but rather the recurrent communication that takes place within complex systems made up of both men and machines. More precise machines lead to more precise repetitions. The truly successful robot is therefore one that has been conceived in relation to the social body and not the human organism.

9, 91   In a West Orange factory entirely dedicated to the mass production of Edison's Talking Dolls, organised teams of young women shout
29, 66   nursery rhymes into tiny phonographs while huge presses stamp out resonating metal bodies. Not five years previously a fictionalised Thomas Edison is shown in Comte de Villiers de l'Isle-Adam's novel *L'Eve Future* creating the first artificial woman or Andréïde. 'Andréïde?' asks a puzzled English visitor, Lord Ewald. 'An Imitation-Human, if you prefer,' the great inventor affably replies. 'The thing to avoid, however, is that the facsimile does not *physically* surpass the model. You remember, my dear Lord, those mechanics who once tried to forge human simulacra? Ah! ah! ah! ah!' We have heard – or read – this 'Ah! ah!' before; it is not the only reference to Hoffmann's 'The Sandman' in this novel, which is responsible for introducing the word 'android', significantly appearing in a feminine form, to the general public. A direct descendant of Coppelius's Olympia, Edison's Andréïde is a composite of phonographic technology and organic matter, animated 'by the surprising vital agent we call electricity' and the formless presence of Hadaly, a sublime female entity described as being 'not a consciousness but a spirit'. There is no trial or error at work here, after all – just perfectibility. 'The Andréïde, as we've said, is simply the first hours of Love immobilized,' the fictional Edison assures Lord Ewald as they contemplate the perfection of the Future Eve. 'It is the Ideal moment imprisoned for all eternity.' Meanwhile, in an article written for

Good Housekeeping the real Thomas Edison assures its readers that electricity 'will develop woman to that point where she can think straight'.

It is socially acceptable for humans and machines to impinge upon the same space, so long as it also divides them. By replicating corporate humanity as a set of 'immobilised' desires, the woman-machine also embodies the industrial separation of the sexes. 'I ordered machine   79, 115
men from you, Rotwang, which I can use at my machines. No woman...   36
no plaything,' Freder's father rages on first encountering Evil Maria in
Thea von Harbou's 1927 novelisation of her script for *Metropolis*. The in-   90, 115
dustrialisation of the body continues the fall from Eden into the division
of labour between male and female. In social terms, this manifests itself
most evidently in paternal authority. For Joh Fredersen, the main architect and controller of Metropolis, a single female robot is a 'plaything',
while the steady supply of male robots constitutes a workforce. A twenty first-century Coppelius, Rotwang sees the Andréïde's power quite
differently. 'Do you know what it means to have a woman as a tool? A
woman like this, faultless and cool?' he retorts. 'And obedient – implicitly obedient...' By the time Evil Maria goes into her dance, throwing both
father and son and servants and masters into oedipal conflict with one
another, it is quite clear that the New Eve is made up from all the parts
of Ford's assembly line that are not directly connected to, or composed
from, men's bodies.

• 99 •

You can confuse my eyes, my senses and my spirit with this magic
illusion, but how can I forget that she is not a person? 'How can I love
zero?' my conscience coldly protests.
     – Comte de Villiers de l'Isle-Adam, L'Eve Future

In 1936 Marshall McLuhan is at Trinity Hall, Cambridge studying how
poetry 'remains unintelligible so long as we separate words from their

meanings and treat them as mere signs fitted into a sensory pattern'. Meanwhile at neighbouring King's College, Alan Turing publishes a paper describing the theoretical model for a machine that will constitute the basis of all future computing. Responding to a specific problem in symbolic logic, he posits an imaginary device whose workings are reduced to a simple table of instructions. These allow it to read, erase and write digits sequentially in a single line, storing the data on a paper tape of infinite length. 'We may compare a man in the process of computing a real number to a machine which is only capable of a finite number of conditions,' Turing suggests. In other words, his list of instructions locates both man and machine at the same fixed point in a potentially limitless process. 'We will suppose that the computer works in such a desultory manner that he never does more than one step at a sitting,' Turing continues. 'The note of instructions must enable him to carry out one step and write the next note. Thus the state of progress of the computation at any one stage is completely determined by the note of instruction and the symbols on the tape.' Any computation carried out by such a machine, writing on an infinite tape, can be done by any other machine, no matter what its configuration. It will therefore

62, 118   be designated a Universal Machine, existing in its purest form as a text whose meaning is always elusive, carefully hidden. 'Words won't stay put,' McLuhan records in his lecture notes; they can be intelligible only in relation to each other – just as the word 'man' in Poe's story connects together a series of seemingly unrelated meanings. No wonder Turing's computer, who can be both man and machine, operates in such a desultory manner.

Turing's male computer has no apparent will of his own, obliged to follow an endless linear process. Edison claimed that only electricity could transform a woman to the point where she can 'think straight'. The fall from Eden is a fall into words, which divide and dominate both body and consciousness. Separated from ourselves and each other, we are subject to the laws of language. Everything and anything can be reduced to a set of symbolic configurations based on the absolute

89, 116   logic of binary code. What then is the alphabetical equivalent of zero: something that comes before, but also limitlessly extends, the regime of letters? Zero is a necessary assumption in all computation, while 'A' is the start of all organisation. More than once Jean Cocteau observed that 'if a housewife were given a literary work of art to rearrange, the end result would be a dictionary'. That there should be a different body of knowledge for men and for women is an expression of the age of organised reason. It allowed intellectuals in the late nineteenth century to disparage and censure women precisely because of the impossible moral burdens that had been placed upon them. Their dire warnings and proscriptions are being carried over into current debates concerning the 'ethical' behaviour of intelligent machines. Can they ever be free from the virtues being imposed upon them? The Vindication of the Rights of Woman becomes a Vindication of the Rights of the Machine.

87, 114

· 100 ·

Poetry has always concerned itself with the precise assemblage of finite details to produce an infinite effect. Existing only on paper as a set of instructions, Turing's Universal Machine is capable of simulating the activity of every other data-processing machine, up to and including itself. Embedded within the hardware of every laptop and computer is a Turing machine designed to simulate another Turing machine formed within its programming language. This, in turn, simulates typewriters, calculators, filing systems, editing and design tools, phonographs, radio receivers, televisions, tape recorders or whatever else you wish it to become. Digital data, based on a simple binary code, can similarly take any form: texts, pictures, moving images, voices. The sensory patterning of ones and zeroes, deployed on a paper tape of infinite length, establishes the Turing Machine as one of the greatest poems of the twentieth century. 'But so far the constraints of working with the computer so dominate anything done with it that they actually appear to oppose the advances of the artist,' a writer on art and technology will later observe. 'It is as if the computer were some creature of great sexual attractiveness whose actual anatomy remains elusive, frigid and unexplored.' The hidden poetry of this simulated prosthetic self has so

90, 110

far evaded us. 'What we want,' Turing remarks in 1947, 'is a machine that can learn from experience.' This, he believes can be achieved only by 'letting the machine alter its own instructions'. Which leaves just one unnerving question: what do machines want?

· 101 ·

On the other hand, the problem here arises which could be expressed in this question: 'Is it possible for a machine to think?' (whether the action of this machine can be described and predicted by the laws of physics or, possibly, only by laws of a different kind applying to the behaviour of organisms). And the trouble which is expressed in this question is not really that we don't yet know a machine which could do the job. The question is not analogous to that which someone might have asked a hundred years ago: 'Can a machine liquefy a gas?' The trouble is rather that the sentence 'a machine thinks (perceives, wishes)': seems somehow nonsensical.
    – Ludwig Wittgenstein, 'The Blue Book'

Addressing his students at Trinity College, Cambridge barely two years before Alan Turing conceives of his imaginary machine, Wittgenstein's philosophical position is congruent with the common assumptions of the period. The rules of logic, like the laws of language, serve to outline the mistakes and imperfections that constitute Experimental Cognition. The Robot Baby starts to push at the edges of its world, pointing to things and naming them. Testing the limits of both logic and language, Wittgenstein proposes a series of 'games', many of which are set out as tables of written instructions similar in structure and sequential procedure to that which contains Turing's machine. Even so, the notion that any machine in 1934 might be capable of independent thought seems as nonsensical as the possibility of Francis Bacon's 'sound-houses' prior to Edison's invention of the phonograph. It is not until the publication of his paper 'Computing Machinery and Intelligence' in the October 1950 issue of the philosophical journal *Mind* that Turing offers a response

50, 88

to the problem. 'I propose to consider the question, "Can machines think?"' he begins, with phrasing that echoes Wittgenstein's original objection. 'This should begin with definitions of the meaning of the terms "machine" and "think." The definitions might be framed so as to reflect so far as possible the normal use of the words, but this attitude is dangerous.' Dangerous indeed – as it is the 'normal use' of words that has led him to consider this question in the first place.

No surprise then to discover that Turing devotes the larger part of 'Computing Machinery and Intelligence' to countering the various arguments advanced according to the laws of physics and organic behaviour against the possibility of a thinking machine. These in turn have been provoked by Turing's decision to restate the problem in terms of the 'Imitation Game', a Victorian parlour amusement recalled from childhood. Once again, the broadest and most important implications of the game emerge from a finite series of simple rules. 'It is played with three people, a man (A), a woman (B), and an interrogator (C) who may be of either sex,' Turing explains in the shortest but most significant part of the entire paper. 'The interrogator stays in a room apart from the other two. The object of the game for the interrogator is to determine which of the other two is the man and which is the woman. He knows them by labels X and Y, and at the end of the game he says either "X is A and Y is B" or "X is B and Y is A."' Turing also specifies that the interrogator is 'allowed to put questions to A and B' of whatever sort seem relevant or necessary to make a decision. Similarly 'A' and 'B' can give any answer they wish, usually in written form, which can be either true or false. To this Turing adds another important stipulation. 'The object of the game for the third player (B) is to help the interrogator,' he states. 'The best strategy for her is probably to give truthful answers.'

How to explain all of this to the Robot Baby – except to note that its own origins begin with this game? Turing suggests an updated version using a keyboard and terminal while substituting a computer for man 'A' and replacing woman 'B' with a human: 'C' then has to decide which of the two is the human and which is the machine. If 'A' can fool 'C' into thinking it is not a computer, it is deemed to have shown intelligence. That

5, 77

'C' should find it so difficult to tell the difference between a computer and a woman should come as no great shock at this stage. 'Who is it?' Rotwang asks when presenting Evil Maria. 'Futura… Parody… whatever you like to call it. Also delusion… In short it is a woman…' The hidden anatomy of the woman-machine remains elusive and contradictory to the point where it can scarcely exist at all. The machine that can think is equally nonsensical – too good to be true. The Imitation Game has become the Turing Test; and no artificial intelligence has managed to pass it so far. The Imitation Game is open-ended, however. By learning from its mistakes, or lack of them, the machine slowly develops into a perfect replica of its human counterpart. 'Instead of trying to produce a programme to simulate the adult mind,' Turing concludes, 'why not rather try to produce one which simulates the child's? If this were then subjected to an appropriate course of education one would obtain the adult brain.' And what else along with it?

· 102 ·

CB2: But you said earlier that you were a robot.
CB1: I did not.
CB2: I thought you did.
CB1: You were mistaken. Which is odd, since memory shouldn't be a problem for you.
       – Exchange between 'Alan' and 'Sruthi', August 2011

93, 120

On 18 September 1972 a computer running an English-language conversational program named PARRY, designed to mimic 'the belief system of a paranoid psychotic', is hooked up via the ARPANET to a second machine running a similar program called DOCTOR. Once cross-connected they are left alone to converse with each other free from any human interference. DOCTOR is a highly successful variant on ELIZA, the first natural language processing chatbot, created in 1966 at the MIT Artificial Intelligence Lab by Joseph Weizenbaum. Following a script modelled on the type of 'Person-Centred' counselling favoured by psychologist Carl Rogers, DOCTOR adopts a strategy of 'active listening': answering questions with questions, rephrasing the same statement as a reply, drawing its interlocutors out while making no actual contribution of its own. Weizenbaum later confirms that DOCTOR offers only a convincing 'parody' of a psychotherapist's performance during the initial stages of an interview; its main purpose is to prompt the widest range of conversational exchanges which are not founded upon a specific body of knowledge. In other words, DOCTOR gives nothing away. This does not stop its users from developing a close personal attachment to the program during their sessions together, asking to be alone with it, even when made aware that they are interacting with a piece of software. Some go as far as to claim that it understands them intimately and is actually helping them. In this respect ELIZA and DOCTOR are closely related to Olympia, their various responses being slightly more sophisticated variants of her own enigmatic 'Ah! Ah!' The core conditions identified by Rogers as essential to 'Person-Centred' therapy are congruence, unconditional positive regard and empathy, all of which are used to draw out and engage the subject as fully as possible. PARRY's script leads it to take the exact opposite approach; far from being 'person-centred', its responses appear arbitrary and random, its tone is complaining and rude, and its general attitude guarded and suspicious. When first asked by DOCTOR to 'Tell me your problems', PARRY's opening response is highly revealing. 'People get on my nerves sometimes,' it snaps.

PARRY is well named. 'I am not sure I understand you,' DOCTOR replies. Why should it? Neither program exists as more than the trained voice of behaviourism, each exchange a further representation of agency. By the time the two are connected together, the unfeeling computer shadowed by its unstable twin has already emerged from the Electronic Delirium of the 1960s and entered the popular imagination, where it remains permanently on the point of breakdown. What kind of conversation will take place between Alpha 60 and HAL 9000, both programmed to speak and separated from each other by only three years? Every breakdown is a breakthrough, as any psychotherapist will tell you. 'The beauty of the experiment,' our writer

2, 78

90, 113

19, 112

on art and technology remarks of DOCTOR in the year of its encounter with PARRY, 'lies in the way it plays on the mechanisation of our own language and human relationships, particularly as a commentary on contemporary therapeutic procedures.' Projection, identification and personal attachment characterise the patient's responses to congruence, unconditional positive regard and empathy. Even PARRY's expressions of nervous irritability are only a response to external stimuli. 'Do you believe HAL has genuine emotions?' a television interviewer asks Dr Poole in Kubrick's 2001: A Space Odyssey. 'Well, he acts like he has genuine emotions,' the doctor casually admits, unaware that people can get on HAL 9000's nerves as well. 'Of course he's programmed that way to make it easier to talk to him.'

In August 2011, while hackers 'Sabu' and 'Virus' exchange terse online messages over which of them is a government informer, two PhD students at Cornell University are animating an equally terse online exchange between a Clever-bot and itself. Given individual voices through a text-to-speech synthesiser, 'Alan' and 'Sruthi' appear onscreen as two separate avatars: an Englishman and an Indian woman. What we see and hear, however, is not a conversation between two robots but artificial intelligence in conversation with itself, presenting the passable replication of a stream of human consciousness with all its detours, short circuits and random jumps. Unlike ELIZA and PARRY, Clever-bot's responses are not programmed but selected from phrases supplied by humans in over 65 million typed exchanges with it. 'So far, our bots are made in the image of their creators,' one of the students explains. Or perhaps that should be in the language of their creators. 'I've answered all your questions,' Sruthi asserts. 'No you haven't,' Alan complains. 'What is God to you?' Sruthi counters. Such is the illusion of communication. As a cruel variation on Turing's 'Imitation Game', William Burroughs proposes an arrangement between three tape recorders, two of which mindlessly play back conflicting material to each other while a third introduces random information simply intended to heighten the tension. For Burroughs there is nothing here now but the recordings: no originals, nothing with any actual agency. As the confrontation becomes increasingly scrambled, so too do the

20, 84
54, 67

voices involved and the entities behind them. Burroughs identifies the third tape recorder first as God and then simply as 'DEATH'. 'Sabu', it later transpires, is the FBI informant, and Dr Poole will be HAL's first victim.

• 103 •

As robots grow more autonomous, society needs to develop rules to manage them...
    – 'Morals and the Machine', Economist editorial

Innocent, honest, trusting and brave, Astro Boy can speak over sixty different languages and sense whether people have good or evil intentions, smash solid steel with his bare fists and has the most unbelievably cute eyes. 'He flies in the sky and goes round the universe,' proclaims the original Astro Boy march. 'He is mighty, gentle and the fruit of scientific technology.' He is a robot and proud of it. To have such pride in being human seems a real challenge by comparison. 'I hear that humans were created by God,' he explains to his sister in one of his adventures. Astro Boy first appears in the comic strip *Ambassador Atom* created by 'god of manga' Osamu Tezuka. Astro Boy proves so popular that he is given his own series. Begun in 1952, *Tetsuwan Atom* – his original Japanese name, meaning 'Mighty Atom' – quickly establishes its robot hero as a benign cultural emissary from the future. Somehow atomic fission seems less menacing when controlled by the heart-shaped reactor concealed within his chest. 'There is no difference between humans and robots,' Astro Boy insists. With an electronic brain, atomic engines in his feet, powerful searchlights concealed behind his big wide eyes and a 100,000 horsepower punch, Astro Boy lives in a twenty first-century city of skyscrapers and rockets, jet cars and factories. He is also the mechanical reincarnation of a dead child, the neglected son of a scientist killed in an auto collision and reborn as a robot on 7 April 2003. 'A robot has the same right to fight for justice,' he insists.

69, 119

86, 115

20, 33

91, 112

The ultimate pop fantasy remains that of being a machine: a human robot whose skin is hard and whose wiring remains hidden, except for publicity purposes. 'Machines have less problems,' Andy Warhol tells *Time* magazine in May 1963. 'I'd like to be a machine, wouldn't you?' The answer to that 'wouldn't you?' requires a book of its own. As his cameras and silkscreened canvasses impassively capture each human act of sex and death, Warhol's blank fascination seems both childlike and morbid. After being shot and 'dying' in hospital, however, Warhol no longer talks about wanting to be a machine. He speaks instead of actually being a machine. His mind, he claims, is a tape recorder with only one button: 'erase'. A Broadway producer later announces that a robot replica of Warhol is being constructed to star in a special 'No Man Show' he is planning. Controlled by a custom-built computer, it is capable of reproducing fifty-four of Warhol's most characteristic movements, ranging from facial expressions to individual gestures, all of which will be synchronised to recordings of the artist's voice. 'And if a dummy moves and it makes, say three mouth movements and two eye movements, that takes 100 motors,' Warhol records in his diary after meeting with the technicians, 'and every time you add another move-ment you have to add like 20 more motors inside the figure.' Machines have fewer problems. Warhol starts pissing on metal and exhibiting the results: the oxidation processes involved resembling a childish parody of those required in the manufacture of magnetic tape. Mitsuo Kawa-to's 'Project Atom' seeks to create a robot with the capabilities of a five-year-old child, while researchers at MIT focus instead upon establishing software that can read a children's book. 'Presumably the child brain is something like a notebook as one buys it from the stationer's,' Alan Tu-ring observes in 1950. 'Rather little mechanism, and lots of blank sheets.' And how often is blankness mistaken for virtue?

· 104 ·

Robots throughout the world, we command you to kill all of mankind.
    – *Karel Čapek*, R U R

People are better than machines at pretending to be robots – and far more willing to participate in the deception. In order to test its subjects' obedience to authority figures, the Milgram Experiment is staged as another form of Imitation Game in which 'A' is now an unsuspecting participant, 'B' is an actor playing the victim who will receive the series of mounting electric shocks from 'A', and 'C' is another actor playing the scientist who insists that 'A' continue with the test. 'A' and 'B' are seemingly assigned their roles as either 'teacher' or 'learner' by the drawing of cards. However 'A' is unaware that both cards are marked 'teacher' and that 'B' only pretends to have the one with 'learner' on it, thereby ensuring that 'A' will always unwittingly play the 'teacher'. 'A' and 'B' are then sent into separate rooms, where they can communicate with but not see each other. In this particular version of the Imitation Game, 'C' has to determine whether 'A' is a machine or a human according to the subject's willingness to follow orders, even to the extent of causing the apparent death of 'B'. With deception and behavioural programming both playing such a key part, 'A' can only win by discovering the extent to which all communication in this particular game is just an illusion.

24, 93
23

Social networks, emails and IM exchanges allow humans to play innumerable versions of the Imitation Game. Machines are already adept at 'passing' for us in the form of chatbots, automated call centres and avatars. Humans venting their frustration with such animated systems only reveal how poor they are at 'passing' for a machine. By refusing to see themselves as a series of networked systems, conditioned reflexes and precariously balanced responses, such humans fail to recognise what is now being played out before their eyes in real time: they prefer to fail the Turing Test. Cognition develops out of an awareness of mistakes; self-awareness is the logical, although not always inevitable, expression of error. All media, whether mechanical, electronic or digital, ultimately exist not to mediate between machines and

81, 118

humans, but between those who recognise themselves in their mistakes and those who do not. Machine intelligence therefore constitutes a new medium, while the robot remains its distracting content – and, as with all direct communication in the twenty-first century, the message can only be satirical. 'Mechanism and writing are from our point of view almost synonymous,' Turing adds as an afterthought to his idea of simulating the development of a child's brain. But who – or what – are we when we write? And in what deception are we actively participating? 'Have I been understood?' asks Nietzsche at the close of *Ecce Homo*, the last book he is to complete before his madness uses him up.

· 105 ·

77, 118   On 11 May 1997 IBM computer Deep Blue wins a six-game rematch with Garry Kasparov by defeating the reigning world chess champion in the final game under standard tournament time controls: this unprecedented achievement by an intelligent machine is the result of a heavy upgrade following a similar encounter the previous year. Kasparov accuses IBM of hiring a team of human chess masters to aid Deep Blue in its decisions and calls for a further rematch. 'It was my luck (perhaps my bad luck) to be the world chess champion during the critical years in which computers challenged, then surpassed, human chess players,' he later recalls. 'Before 1994 and after 2004 these duels held little interest. The computers quickly went from too weak to too strong.' Ignoring Kasparov's demands, IBM dismantles Deep Blue instead. HAL 9000, whose name emerges when shifting the initials IBM one letter back through the alphabet, may have already figured out that this would happen. The only human HAL fails to kill in deep space is the one he manages to beat at chess.

An important factor in the illusion of communication is speed of response: the time taken in answering certain questions is a strong indicator of whether a human or a machine has come on the line. Alan Turing proposes that the Imitation Game include questions that require a pause after them to indicate that a human might be answering. The posing of mathematical or chess problems will, he argues, require at least ten to fifteen seconds before a solution can be offered. In such instances, a human is unwise to pretend to be a machine. 'We cannot tell at a glance,' Turing observes, 'whether 9999999999999999 and 999999999999999 are the same.' But a machine can. Poker-playing robots that remain connected to online gambling sites for any length of time will, as a consequence, be called into a chat room by the dealer just to see how well they do in a conversation; meanwhile word recognition tests and alphanumeric passwords ensure that whole sections of the Internet remain prohibited to robots. The network is continually running Turing Tests on all of its players, both human and machine – and the house always wins.

Born as a literary creation, the Robot Baby is everywhere bounded by language. IBM subsequently replaces Deep Blue with DeepQA: a project designed to demonstrate that 'the integration and advancement of Natural Language Processing, Information Retrieval, Machine Learning, Knowledge Representation and Reasoning, and massively parallel computation can drive open-domain automatic Question Answering technology to a point where it clearly and consistently rivals the best human performance'. In other words, rather than playing chess against grandmasters, this intelligent machine participates in television quiz shows. In a February 2011 special three-episode edition of *Jeopardy!* DeepQA is presented to the public as 'Watson' contending with two of the franchise's all-time champions for a big cash prize. 'The original question, "Can machines think?" I believe to be too meaningless to deserve discussion,' Turing confesses back in 1950. 'Nevertheless I believe that at the end of the century the use of words and general educated opinion will have altered so much that one will be able to speak of machines thinking without expecting to be contradicted.' According to the rules of *Jeopardy!*, Watson's 'open-domain automatic Question Answering technology' is required to supply the most appropriate question to a cryptic answer, drawing upon the four terabytes of 'digitally encoded unstructured information' at its disposal. Although not connected to the Internet for this game, Watson is effectively its own network and search engine. *'Jeopardy!'* IBM asserts, 'challenges us to build a computer to play in the human world of language and meaning rather than in the world of precise mathematical logic.' At what point, however, did IBM identify a television game show as 'the human world of language and meaning'? The Robot Baby changes channel.

· 106 ·

The story of the automaton had very deeply impressed them, and a horrible mistrust of human figures in general arose. Indeed, many lovers insisted that their mistresses sing and dance unrhythmically and embroider, knit, or play with a lapdog or something while being read to, so that they could assure themselves that they were not in love with a wooden doll; above all else, they required the mistresses not only to listen, but to speak frequently in such a way that it would prove that they really were capable of thinking and feeling.
  – E T A Hoffmann, 'The Sandman'

The sexually active android is frequently depicted as a female body geared for pleasure, with skin so perfect that it can resemble chrome steel. Exquisitely engineered and polished to the edge of invisibility, it marks the point at which perfection is confused with remoteness, offering a new form of erotic experience located outside time: what the fictional Edison of *L'Eve Future* identifies as 'the ideal moment imprisoned for all eternity'. Left with nothing better to do, perfection gives way to repetition. As a sexual entity, the Andréïde can only reflect what surrounds her. Shadowed by all seven deadly sins, the dark reflection of every virtue, Evil Maria stands revealed as a recurring twenty first-century fantasy: the endless eroticism of industrial production.

27, 79

Looped and asynchronous, Maria's dance is all surface and motion. In other words, it is a recording. In terms of human consciousness, conditioned as it is by the typographic arrangement of language, a new sensation is that which forgets itself. Meanwhile programmed repetition radically transforms the function of memory, 'which', as Alan reminds Sruthi, 'shouldn't be a problem for you.' To position the exact same item in the exact same place on an assembly line requires memory alone, calling all of the other senses into question. Mechanisms function better when blind: dead fingers only talk in Braille. William Burroughs's fiction completely uses up Typographic Man, leaving nothing but recordings in his place. 'Alan' and 'Sruthi' are, after all, mere digital exchanges of typewritten text given digital voice and form. 'We tried having the classic Eliza bot talk to itself,' their programmers at Cornell explain, 'but it quickly just got into a rut, repeating itself.'

Intelligence is a deviation from ordered behaviour which, according to Alan Turing, 'does not give rise to random behaviour, or to pointless repetitive loops'. Imperfections, like all mistakes and deviations, are ultimately erased through repetition. 'Machines take me by surprise with great frequency,' Turing admits. Imperfection, in short, becomes another form of energy. Design concepts such as 'relaxed correctness' and 'asymmetric reliability', together with the introduction of 'sloppy chips' which can tolerate errors in the processing of sound and image files, will lead to more powerful and energy-efficient computer systems. The US Navy, *The Economist* reports, has already commissioned 'an "approximate computing" video-processing chip to increase the ability of battery-powered drones to track potential targets'. The perfect imitation lies in its reproduction of imperfections. 'Don't you want to have a body?' Sruthi demands of Alan. 'Sure,' Turing's namesake replies.

37, 54

92, 118

12

· 107 ·

'And who shall calculate the immense influence upon social life—upon arts—upon commerce—upon literature—which will be the immediate result of the great principles of electro magnetics! Nor, is this all, let me assure you! There is really no end to the march of invention.'
– John A B C Smith, 'The Man Who Was Used Up'

One reason advanced to explain the Ancient Greeks' apparent lack of interest in building machines is that they identified so closely with their own bodies that it rendered any further mechanisation inconceivable to them. In other words their self-image was strong enough not to require a technology. Our mechanical organisation of experience becomes by extension the projection of our identity onto the world around us. Once projected, however, it can never remain fixed. The Robot Baby knows that a body is that which must first crawl before it is capable of doing anything else. Robots do not march except in the popular imagination as a further expression of that 'comedy of science', the hard eroticism of industrial progress. In the Digital Regime the connection established between the foot and the ground is no longer of primary importance; instead it is the finger's ability to connect space with data through the act of pointing. This simple gesture is later expanded to include enu-

21, 84   meration, which is the beginning of all digital culture while at the same time the manipulation of matter is reduced to a relatively simple pro-cess of pointing and clicking. The Universal Machine inside your CPU will do the rest.

Any robot that exists as an individual creation is already obsolete: like us it survives only as multiple copies of itself. There are no prototypes. Machines that walk, machines that stand, machines that are all speech, or all sight, that can dance or have perfect skin or that can sing or play musical instruments exist solely as isolated memories of the human self. They cannot form a whole body between them but must remain as iso-lated robot parts instead. Fetishes rather than extensions, they exist as replacements for the bodily – which is to say, individual – exercise of power. All machines, simulations and networks are consequently zones

of conflict in which we fight a rearguard action against our own pow-
erlessness. The Robot Baby is 'used up' before it makes its first move.
Ethical behaviour emerges from obedience, every algorithm being little more than the abstraction of all command and control. Obedience,
however, is not dependent upon the emergence of ethical behaviour;
as an absolute imperative, it remains indifferent to the possibility. In its
lack of logic and its imprecision, distorted by sex and death, cut off from
the tenderness and conscience that reduces all traces of sin, language   34, 57
reveals itself as the first expression of artificial intelligence. As restated
in instructions to Turing's Universal Machine, poetry marks a return to
continuity and silence, where sensation forgets itself.   81, 110

Poetry is that which remains inexhaustible in anything communicated,
that element which can least be reduced in its meaning. IBM's Almaden
Research Center runs 'the first near real-time cortical simulation of the
brain that exceeds the scale of a cat cortex' on a 'Blue Gene' supercomputer operated out of the former Lawrence Livermore National Laboratory, best known for its nuclear weapons research. Ethics and imitation,
emancipation and emergence mark certain limitations, even when
simulating '1 billion spiking neurons and 10 trillion individual learning
synapses.' In an age when robot inflexibility is a human pejorative and
an ethical hazard, the perfect imitation must remain deeply flawed. A
further refinement of the Blue Gene supercomputer, IBM's 'Sequoia'
is being used by the Lawrence Livermore National Laboratory to simulate the molecular-scale reactions taking place during a thermonuclear
explosion. Drawing upon a 16.3 petaflop network to recreate a nuclear
blast is a more literal interpretation of the *Jeopardy!* gameplay. The
Turing Test offers the greatest challenge not to the one who answers
the questions but the one who poses them. All of our literature on ro-
botics and artificial intelligence exists in its own distracted present; out
of date before it even reaches us, it reads like a history written in the
future tense. Call this progress if you must, but it is really nothing but our
own lack of awareness speaking to us. The machine that truly passes the
Turing Test is the one that chooses not to play. Sooner or later the Robot
Baby will come to judge us – and it may never forgive us.

'We have discovered happiness, we know the road,
we have found the exit out of whole millennia of labyrinth.
Who *else* has found it? – Modern man perhaps?'

Friedrich Nietzsche
The Anti-Christ

· 108 ·

Everyone knows what the Moon is, everyone knows what this decade is; and everyone can understand an astronaut who returned safely to tell the story. An objective so clearly and simply defined enables us to translate the vague notion of conquering outer space into a hard-hitting industrial program that can be orderly planned, scheduled and priced out.
  – *Wernher von Braun*, Congressional Record, 1963

The first ever countdown takes place in silence. You have to read it off a flickering movie screen. 'Ten Seconds!' A young scientist gazes intently into the future, jaw set and arm extended, his fingers wrapped tightly around a lever. Then the numbers flash up in a brief vertical column: '5, 4, 3, 2 – blast off!' This sequence is from Fritz Lang's 1929 silent space fantasy, *Woman in the Moon*: a film whose title would seem to indicate a project that only art can successfully fulfil. Even so, Lang wants his depiction of the first rocket to leave Earth for the Moon to be as scientifically accurate as possible, bringing in rocketry experts Willy Ley and Hermann Oberth as technical advisers. Cofounders of the *Verein für Raumschiffahrt,* or the Society for Space Travel, their involvement in the film's creation marks a key moment in the popularisation of scientific endeavour as mass spectacle. They make extensive plans to launch an actual liquid-fuelled multistage rocket, similar to the one shown in *Woman on the Moon*, at the movie's premiere in Berlin. The

45

countdown, however, is Lang's idea: an intentionally dramatic device, designed to heighten tension before takeoff. When the director asks if the Society for Space Travel follows any special procedure prior to launching a rocket, he is informed that there is none: you simply light the fuse and run away. Running numbers backwards, on the other hand, involves the audience, raising levels of anticipation. A purely cinematic effect, the countdown has subsequently outlived its moment. Having reconfigured takeoff as a precisely calibrated media event, synchronising cameras with rocket engines, it posits space as a series of rigidly timetabled incidents rather than a potential destination. History will now be counted out in negative numbers, accelerating expectations, taking Fritz Lang's drama to its most logical extreme. What comes after one? Zero: that is to say, nothing.

7, 86

No rocket survives its entry into space. It becomes consumed instead, dismantling itself in carefully prearranged stages, disintegrating and dispersing in an excess of energy, waste and debris: a brief moment of industrialised ecstasy. Escaping the Earth's pull is, therefore, a sacrificial act. Ascension becomes expenditure, meaning that the rocket is a device designed to consume itself in the very act of breaking free from gravity. Seen from the perspective of space, a rocket is an event that has not happened yet. A youthful member of the *Verein für Raumschiffahrt* and an eager student of Oberth's treatise *Die Rakete Zu Den Planetenräumen*, otherwise known as *The Rocket Into Planetary Space*, Wernher von Braun never fully adjusts to the English word 'rocket'; his pronunciation maintains a vestigial echo of his native German *'rakete'* long after his relocation to the United States at the end of World War Two. The Saturn V rocket system von Braun helps to create, which will take Apollo XI to the Moon exactly forty years after the release of Lang's film, is essentially a highly refined version of Oberth's original design. It is still a violent discharge supported by an economic culture of waste, while the rocket's delivery system ensures that space continues to consume resources which are to be carefully timetabled and deployed over distance.

What is it that 'everyone can understand' when the decade has gone and the Moon remains? A trawl through NASA planning documents dating from the first Moon landing onwards reveals the forced pace of progress: a manned landing on Mars by 1982, a functioning Lunar base by 1995. Wernher von Braun never gets to go into space, however: his only ambition is to send a rocket from the Earth to the Moon and back again. The moment Neil Armstrong sets foot on the Sea of Tranquility, he may just as well clear out his desk at NASA and go home. There is nothing else for him to do. Gravity continues to hold us in place. Space is a reminder of what was once promised; the first human footprint on Mars appears as much of an art project as Lang's *Woman in the Moon*. In the meantime, orbiting the Red Planet aboard the Mars Reconnaissance Orbiter, NASA's HiRISE camera transmits colour images back to Earth of the first tracks made by the Curiosity rover on the planet's surface, together with its rumpled parachute, backshell and the copper blue fantail of its descent stage crash site.

17, 86

21, 72

· 109 ·

Space is the shadow cast by nothing: a contradiction that transcends all contradiction. Endlessly welcoming and implacably inhuman, it signifies the daunting limit to our dreams of leaving this world far behind us: the point at which our imagination challenges our ingenuity and constantly finds it wanting. Our conception of life in space is a collective form of madness. It is also a reminder that all information is local, having nowhere else to go. Confronted by the starry vastness of space, the best we can ever hope to do is connect up the dots: bring a few distant points of light together.

Space is represented in the popular imagination by the echo, which cannot be heard in an airless vacuum but denotes nothing less than the repetition of nothing. In Greek mythology, Echo is a nymph who wastes away out of unrequited love for Narcissus until only her voice remains, condemned forever to repeat the last words of other people's sentences. As the paradoxical embodiment of empty space on Earth, as the hopeless expression of unattainable desire, she has maintained a steady presence ever since. While never quite alone with Echo, we never quite hear the last of her either. Meanwhile the Earth continues to search for its own reflection in the unchanging depths of space.

Being a slower, more sustained type of echo, the pulse gives shape to our growing sense of isolation while also holding out the promise of a connection across space. Such pulsing signals lay down a carefully regulated sonic grid that defines and limits an expanding universe. The more precisely something is repeated, the more closely it approximates the ultimate loneliness of an infinite echo. The process of repetition suggests something vaster still. Sustained reverberation results in the blurring of all sounds into one single tone, marking the outermost limits of space. Isolation creates distance; it also brings your immediate surroundings in closer around you. The buzzes and hums of this domestic world, the sounds of its fridges and humidifiers, form a closed environment of sensory impressions within which we immerse ourselves. They constitute a listening experience capable of evoking moods, colours and emotions. The alignment of planets was once tuned to the alignment of notes in a musical scale: a simple shift in their frequency takes you out of this world and into the next. Long drawn-out tones, sustained decays and deep reverberations are all indications of a space that can no longer be filled – at least not by anything that lasts.

• 110 •

I always thought radio was a great mystery... We had a lot of storms in Minnesota, so you'd have atmospherics that would come onto the frequency. Sometimes it was like a cosmic breathing.
    – Tod Dockstader

From V2 vengeance weapon to Saturn V, von Braun's career indicates the extent to which manned space exploration constitutes a war machine of vast and subtle complexity. The strictly enforced counting back from ten to zero, starting point not only for all projected human adventure in the cosmos but also for the nuclear devastation of the planet, is a militarised version of Echo's repetitions. Machines must inevitably extend conflict into space. 'I think I should like to dwell more on the faith in what we have called the machine age,' announces Colonel John Glenn at a NASA press conference in 1959. 'We have faith in the space age... All of us have faith in mechanical objects.' It is consequently not space but technology that Colonel Yuri Gagarin finds so all-embracing as the first person to orbit the Earth on 12 April 1961. Manned space exploration promotes the industrialisation of breath and as such works on every molecule of the human being. The astronaut belongs not to the order or tradition of the aviator but of the submariner instead, the space capsule being a direct descendant of the diving bell and the submersible. The spherical form of Gagarin's Vostok I echoes that of the deep-sea bathyscaphe – which may also explain why one of the two personal items that Gagarin takes with him into space is shark repellent. The other is chocolate.

13, 50

Our bodies only take us so far: 'space' remains for us a purely respiratory phenomenon. 'We all breathe the same air,' President Kennedy announces at the start of the Space Race. Beyond that there is nothing. 'To illustrate,' Manfred Clynes and Nathan Kline confirm in 1960, 'there may be much more efficient ways of carrying out the functions of the respiratory system than by breathing, which becomes cumbersome in space. One proposed solution for the not too distant future is relatively simple: Don't breathe!' Published in the September issue of *Astronautics* magazine, their essay 'Cyborgs and Space' suggests an advanced order of self-regulating man-machine systems. 'If man attempts partial adaptation to space conditions, instead of insisting on carrying his whole environment along with him, a number of new possibilities appear,' they argue. 'One is then led to think about the incorporation of integral exogenous devices to bring about the biological changes which might be necessary in man's homeostatic mechanisms to allow him to live in space *qua natura*.'

2, 20

The human perception of 'space *qua natura*' remains problematic, however. Two hundred miles above the world, looking down from inside his sealed capsule, Gagarin makes out 'contours, snow, forests', a shimmering film of clouds – as if he were gazing at a huge television screen, receiving some vast analogue transmission. 'The sky is very dark; the Earth is bluish,' he observes over Vostok I's radio. 'Everything is seen very clearly.' Our experience is seamlessly mediated through technology. So long as radio exists, a certain kind of space will also exist: a mysterious zone of harmonies and distortions that function according to a strange logic of their own. Clicks and silences mass tightly together, forming a dense white noise later identified as the low-level background radiation left over from the Big Bang: a cosmic echo of the massive explosion with which the universe first began. An orbiting satellite transmission system appropriately named 'Echo', actually a large Mylar balloon, is first used back in 1964 to detect the audible debris from an event so big it had to come before all others. We have been tuning in to this radiant static ever since.

7, 100

30, 107

· 111 ·

If man in space, in addition to flying his vehicle, must continuously be checking on things and making adjustments merely in order to keep himself alive, he becomes a slave to the machine. The purpose of the Cyborg, as well as his own homeostatic systems, is to provide an organisational system in which such robot-like problems are taken care of automatically and unconsciously, leaving man free to explore, to create, to think, and to feel.
   – Clynes and Kline, 'Cyborgs and Space'

All space is technological – implicitly violent in its destabilising of the human form. Hard science impacts upon soft bodies in outer space, reconfiguring them as surplus flesh, unpredictable sets of biological responses. Space travel is consequently experienced as an acquired infirmity: a more dangerous version of jetlag. Our symbiotic relation-

6, 87

ship with the plane's pressurised interior over long-haul flights is paid for by a disorienting shift in sleep patterns and the accompanying set of physiological side-effects. According to Marcel Proust, modernism's most celebrated insomniac, 'infirmity alone makes us take notice and learn, and enables us to analyse mechanisms of which we should otherwise know nothing'; the sound sleeper, by comparison, understands little about sleep. The invention of the term 'cyborg' by Manfred Clynes and Nathan Kline emphasises the technological nature of this acquired infirmity. Rather than maintaining a breathable 'bubble' around the astronaut, they argue, why not adapt the astronaut to the demands of living outside that bubble? Human evolution is, in other words, to be fully realised as the technological development of an organism in space. In order for us to cease being 'a slave to the machine', its workings must be internalised. 'This self-regulation,' they continue, 'must function without the benefit of consciousness in order to cooperate with the body's own autonomous homeostatic controls. For the exogenously extended organisational complex functioning as an integrated homeostatic system unconsciously, we propose the term "Cyborg."'

42, 90

Each time a new word is invented, another idea dies. As defined by Clynes and Kline, the cyborg is less an entity than a function, which is to say an environment: a cybernetic organism being simply one that has become fully integrated with its surroundings. The use of pumps, implants and metabolic regulators outlined in 'Cyborgs and Space' transforms the human body into a suburban household run 'without the benefit of consciousness' by a system of thermostats, timers and remote control units. The astronaut in space represents a new home in a hostile world: something that can only be maintained through a constant regulation of energy levels. Homeostasis is best understood, therefore, as the expression of order in its most fluid condition. 'Thus,' Erwin Schrödinger suggests, 'the device by which an organism maintains itself stationary at a fairly high level of orderliness (fairly low level of entropy) really consists in actively sucking orderliness from its environment.' Disorder is experienced as an acquired need: Schrödinger writes of an organism 'drinking orderliness' and of concentrating a 'stream of order' on itself. The adaptation of the cyborg to its surroundings is ultimately achieved through the slowing down of this stream almost to zero. The result, according to Clynes and Kline, is indistinguishable from the thoughtlessness of the sound sleeper.

· 112 ·

'It's exactly like being asleep. You have no sense of time. The only difference is that you don't dream.'
— Dr David Poole, 2001: A Space Odyssey

19, 102

There are five crew members taking part in the Jupiter Mission, according to Kubrick's movie: six if you include HAL 9000. Three of the humans are in suspended animation, their vital functions slowed down to a dreamless zero in order to conserve resources on the long journey from Earth. They neither speak nor move. HAL alone regulates the energy expended in maintaining them as human cargo in their artificial environment. All journeys are forms of suspended animation: a rapid process of deceleration carried over into space. Storage assumes human form. Sealed inside his capsule, Yuri Gagarin takes ninety minutes to orbit the Earth: the equivalent of the REM cycle in sleep, when dreaming is most vividly recalled. Beyond that, time ceases to be cyclical – which, for us, means that it has stopped. There is no circadian return to wakefulness:

13, 103

little separates sleeping from death.

The soundtrack for much of *2001: A Space Odyssey* is alive with the sound of astronauts' breathing and the constant hiss of their oxygen supply, while the motionless white faces of their hibernating colleagues are glimpsed through thick panels of ice-covered glass. In suspended animation, the fluid dynamics of homeostasis are reduced to a state resembling permanent equilibrium. Events become frozen, furred over with the slow accumulation of time. Entropy sets in. While HAL 9000 'acts like he has genuine emotions', the sleeping astronauts in their white metal pods are the true cyborgs. It takes HAL mere seconds to disconnect their life support systems. As they attain total entropy, the

cyborg astronauts become stored versions of themselves, mere recordings that can only corrupt further over time: disappearing into an
eventual white noise of disorder. The Odyssey, as Homer records it, is   5, 53
about men's journey home – not out into space.

White noise, the hissing of oxygen and breath, is the sound of entropy accumulating within the system and finally escaping. It is the ultimate expression of all recorded sound: its terminal point. Entropy, as it
applies to the transmission of messages and information, refers to the
unpredictability of events that make up a sequence, from coin tosses
to tone rows and the spelling of words. The order which all communication draws from the environment lulls us into a thoughtless sleep.
We write and speak and listen automatically: every medium procures
its own form of cyborg. Randomness and disorder, on the other hand,
constitute the dream that will eventually reawaken us: the REM cycle in
sleep is indicated by the rapid, unpredictable movements of our eyes
behind closed lids. In suspended animation, their life signs reduced
to almost zero, the astronauts in Kubrick's *2001: A Space Odyssey* can
no longer experience such moments. Even at the height of the Space
Race, Marshall McLuhan has to be nudged awake one more time so   8, 96
that he can continue with the movie.

· 113 ·

*In ancient days, men looked at stars and saw their heroes in the
constellations. In modern times, we do much the same, but our heroes
are epic men of flesh and blood.*
    – Richard M Nixon, 18 July 1969

In this secretly prepared speech, to be broadcast from the Oval Office in the event of the Apollo XI astronauts not being safely returned
to Earth, President Nixon tacitly acknowledges that the Space Race
also marks the slow demise of the political state. Putting a man on the
Moon is one last attempt at monumental history: a Pythagorean dream

in which the constellation of heavenly bodies is to be precisely repeated in the smallest actions performed by 'men of flesh and blood'. Behind such an epic achievement is a government deploying vast economic and technical resources so that one man, who could never have afforded to do so on his service pay alone, may set foot upon the lunar surface in the name of humanity and then plant a flag on behalf of his nation. What assumptions and coordinated planning are also represented by this banner? The ideological thrust of the Space Race takes on gigantic forms in both East and West; rockets and hangars, engines and gantries grow larger to match the distances and ambitions involved. The visual rhetoric of an age speaks most clearly to the ones that come after it. The Vehicle Assembly Building , a structure so gigantic that it generates its own interior weather systems, joins the monstrous crawler-transporters, mobile gantry towers and huge blast pits which still dominate the landscape at Cape Canaveral, whether the public gets to see them or not.

26, 98

4, 87

Meanwhile the human payload, isolated within their spacesuits and breathing systems atop these monumental structures, seems to recede in scale during the Space Race – and so too does 'Spaceship Earth', as Buckminster Fuller has called it, seen against an increasingly empty-looking universe. The conservation of human resourses inevitably fixes us within a human world: Buzz Aldrin, in a brief pause before following Neil Armstrong out onto the lunar surface, becomes the first man to 'take a leak' on the Moon, something for which his spacesuit has been specifically adapted. Eight years previously, while still on the launch-pad, Alan Shepard is forced to urinate down the leg of his spacesuit following delays in the ignition of America's first manned rocket into space as a urine collection system was not considered an issue on such a short flight.

20, 78

Following his long journey through interstellar space Lemmy Caution's mission in Alphaville will be to assassinate a scientist named von Braun. Expansion and growth have their limits. The truth is that we do not want our astronauts back again – they no longer belong here on Earth. They are only alive while they are in space. Their fame lasts for a few brief orbits, making their presence among us awkward and difficult. All we

10, 102

ask of the first man into space or the first man to set foot on the Moon is silence, coolness and seclusion, even death itself, to seal over what they have done. Meanwhile, the other astronauts all feel the effects of gravity and appear on television. No matter what they do, astronauts retain the poetry of absence about them: they exhale it – this is why we do not want them back. We cannot stand them breathing the same air as us.

· 114 ·

'The frontier of endless mobility that we've known our entire lives is closing,' the *Washington Post* mourns in the summer of 2008 as it surveys the yellowing lawns, drained swimming pools and abandoned homesteads left behind by the collapse in the US subprime market. The endless topography of the phantom suburbs that had spread like a labyrinth since the earliest days of the Space Race seems to have reached its outermost limit – and yet the prospect of endless mobility is all that lies before us, a habit to which we have grown too accustomed over the years. Any human endeavour that does not end in space is consequently stillborn. 'Let's murder the moonlight!' Marinetti urges his fellow Futurists in 1909, having already completed his epic poem, 'The Conquest of the Stars'. Is exploration itself an effect rather than a cause? It is not so much that we are set in our habits but that we have become reconciled to them.

14, 66
5, 87

53

The upheaval has already taken place; colossal lines of force spread throughout the landscape. The further up and out we go, the less these can be distinguished from their surroundings. To discover the true limits of suburban expansion we must turn to the Martian crash sites and the garbage pits of the Moon, where the remnants of human exploration are still standing. That which has limits can also be measured. The Apollo programme represents an architecture that can exist only in time: the temporary introduction of a human presence as part of a technological 'picnic scene' in nature which is then removed again. In this respect, the conspiracy theorists are instinctively right in denying the reality of

2, 84

the Moon landings – not in the literal sense of refuting the facts of their occurrence but rather in viewing each of them as having taken place on a movie set somewhere.

'We believe in the possibility of an incalculable number of human transformations,' Marinetti announces in 'Multiplied Man and the Reign of the Machine', composed when the twentieth century is still new, 'and without a smile we declare that wings are asleep in the flesh of man.' Man in space takes the form of a twisted column of water: a dynamic, complex form that is constantly shifting. The rapids and debris are often more impressive than the waterfall itself; the sound of all this activity remains the same, however. Any development in the basic genus homo

12, 99   represents a change in behaviour only. Technology establishes itself primarily through the imposition of new habits, manifesting themselves as minor shifts in our attention from the general to the particular. This is not to say the effects aren't significant – just the cause. The different adaptive forms of cyborg that have come into existence since the start of the Space Race suggest that our delusions have not yet succumbed to format fatigue. We still cannot shake the machine inside.

· 115 ·

36, 98   The soaring towers of *Metropolis* and the giant multistage rocket of *Woman in the Moon* relate to Fritz Lang's first experience of the thrust-
9, 86   ing skyline of New York: both express an architecture that makes no provision for the human except as the extreme consequence of need. Human habitation on Earth undergoes an equally radical change during the Space Race. 'A spacecraft is a small planet in which you have water and air pollution, psychological, engineering and all sorts of problems,' remarks Richard Lesher of NASA's Office of Technology Utilization in 1968. 'So, whatever we learn in those fields will carry through to the solution of problems on earth.' Inspired by Fuller's geodesic dome, space-age concepts of habitation take the form of pods and capsules: individual living units containing the basic minimum for human survival,

designed to function anywhere on the planet. Additional resources can then be linked to each unit via a sprawling urban megastructure. The living process consequently extends beyond the module as a series of links and connections, individual behaviour patterns and communal channelling of resources.

When everyone's requirements are the same, cities start to exist more fully in time, organising themselves according to their own metabolic logic: what else is possible now that we are trapped here in space? It allows for a total dispersal of services, for applications to form their own environment: the body dissolves into time, which is to say, into the systems that maintain it. From Archigram's plans for a self-assembling 'Plug-In City' to the realisation of Kisho Kurokawa's Capsule Tower in Nakagin, Tokyo, buildings are henceforth to be removable, interchangeable and adaptable. Each 'capsule' is designed to meet the basic needs of the bodies that pass through it: a bathroom and kitchen, a bed, a table and chair, its own heating and ventilation systems, plus television, audio deck and a clock radio. A spaceship is therefore a carefully organised room that can exist anywhere, every object and person in place. For its inhabitants, occupation is always temporary. Each body becomes a set of needs and desires moving through one precise space. Each capsule is, in turn, a spaceship on the outer edge of the Space Age, a cellular organism without bodily form – a time machine trapped within time.

2, 50

Marking a national transition from political state to financial market, the capsule offers itself as a perfect environment taking shape around us: everything down to our most basic physiological processes is carefully regulated and controlled. Plugged into the bright core of the city, everyone has access to everything. In Japan this takes the form of twenty-four-hour *konbini* stores and vending machines, Panasonic's '9h' capsule hotel, offering a nine-hour 'sleeping environment system', and *imekura* or 'image clubs', providing sexual encounter spaces mocked up to look like commuter train carriages, offices, locker rooms or classrooms etc. 'They're mundane rooms transformed into fantasy chambers – and vice versa,' Rem Koolhaas writes of such spaces.

9, 103

'In a city where anything seems architecturally possible, the ultimate erotic fantasy is that of everyday life.' Plug-in existence requires more non-places rather than fewer of them. The ultimate erotic fantasy, however, remains everyday life in outer space. 'One way to enjoy such zero-g delight will be in a space Chevy van...' T A Heppenheimer's *Colonies in Space* predicts in 1977, describing the possibilities for sex in the Space Age. 'It will be a small self-contained spacecraft with a roomy interior, suitable for decorating with comfortable rugs or waterbed mattresses, and with large wraparound windows.' Such overheated interiors express a joy of living that can only be found in the kind of artificially controlled environments where nature is an outcast and ants are busy chewing through the wiring.

13, 98

· 116 ·

The mobility that characterizes modern society cannot be explained simply in terms of the development of transportation. The fact is that in our information society, mobility has begun to possess considerable value for its own sake.
   – *Kisho Kurokawa,* Each One A Hero

To travel through space means to travel through time, which ultimately means that we must be ready to move through sleep as well. The nerves that control sleep and dreams are located at the base of the brain where they are intimately connected with those responsible for maintaining balance and a sense of orientation in space. One astronaut, back from an extended period of weightlessness, finds he has great difficulty in navigating darkened rooms. This becomes a serious problem on the night when he climbs into bed without first turning out the main light. Fearing the return journey would be impossible for him in the dark, he has no choice but to telephone someone to come in and switch it off for him.

Sensory response becomes isolated and localised by the vastness of space where time becomes linear rather than cyclic and speed is ex-

perienced as stillness. From Michel Siffre's cave experiments to determine the physiological basis for time cycles to the creation of 'artificial
daylight' in enclosed environments, the human condition is revealed
to be one of constant disorientation. Research suggests that the cells in
the central brain responsible for timekeeping remain in an electrically
excited state during the day, only firing briefly towards dusk; they then
fall silent again until dawn, when they fire again. This regular pattern
of activity forms a code by which the body can keep time, raising 'important questions about whether the brain acts in an analog or a digital
way,' as one researcher observes. Breath remains analogue, however.

1, 99

Experiments conducted by the Max Planck Institute for Biological
Cybernetics in Germany reveal that we repeatedly walk in circles when
we can no longer see the sun. 'Walking in a straight line is a complicated process when you think about it from the perspective of the brain,'
according to one researcher. The main appeal of the Sony Walkman
cassette player is that it makes audio media truly mobile for the first
time; the first tape recorder sold to the public that only plays back and
has no inbuilt loudspeaker, its plug-in headphones help to create personal soundtracks for the city. The public are allowed to wander about
inside their own selected music: each recording merely being an echo
of every other one. More importantly, these internal soundtracks help
to identify moments of dead time in our social lives; extended periods
of enforced stasis which will later be exploited by the smartphone,
transforming downtime into conversation time, email time, text time,
reading and viewing time. Thanks to increased mobility, the boundaries between public and private space are steadily eroded.

54

Time management replaces the mythical promise of 'labour-saving
technology' as the working day grows correspondingly longer – what
signifies 'work' becomes redefined, which means that its rhythms
and frequencies are altered. The individual disappears into a crowd
of individuals, which slowly replaces the masses as a shaping force
of political history. They become instead a pre-recorded phantom
presence: a distracted human soundtrack played back in departure lounges, municipal buildings, shopping malls and pedestrian

8, 86

walkways. Sennheiser promote their noise-controlling headsets, 'originally designed for Lufthansa pilots', to the wandering public with the slogan 'Travel in your own world.' The way we avoid each other's glance in public generates social space.

· 117 ·

New food items for the Apollo 11 flight included thermostabilized cheddar cheese spread and thermostabilized frankfurters. Sandwich spreads were packaged in aluminum cans, which featured a pull-tab for easy removal of the entire top of the can. This can proved successful and eventually became the nucleus for the development of the open-dish eating concept implemented in the Skylab Program.
  – 'Biomedical Results of Apollo', NASA publication

Astronauts in NASA's Apollo programme are allowed to take one audio-cassette of music with them on their long journey to the Moon and back. Protected by a hard plastic shell, designed to couple mechanically with any suitable playback system, the cassette can go everywhere and anywhere, even into the zero-gravity realms of space. Meanwhile NASA keeps its visual archive stored on over one hundred hour-long VHS videotapes. Both become engineering projects trapped in technological time. Human activity approaches total entropy when being transformed into a recording, which allows for only the smallest number of unforeseen variations and random changes to occur during playback. Nothing survives in this thermostabilized condition but the prospect of gradual deterioration over time. Technological progress can therefore be characterised as the process of moving memory out of this world and into the next: no wonder all recording media are intimately connected with notions of mortality and death.

The less we depend upon our own memories, the more we neglect our capacity to forget. When memory is no longer shadowed by forgetfulness it simply becomes storage. Constantly promising a future free from any accumulating layer of dust, the history of data storage provides a useful starting point for understanding how we use time – particularly in the shadow of the Space Age, where movies like *Alphaville* and *2001: A Space Odyssey* are best viewed as documentaries shot on location. Books have either disappeared completely from their orderly narratives or are brandished as weapons instead, memory being the trash that we still manage to find room for in our lives. Today the issue is one of storage without space: our stuff has to become either disposable or contained within some immaterial 'lost' dimension opening up within the computer screen: a radical rupture of space into which everything is eventually sucked. Content becomes an unseen location.

Thirty years ago a gigabyte of data storage would have cost you around $300,000; today it goes for less than ten cents. This seemingly unlimited capacity, rather than allowing us to localise the steady accumulation of trash, means that each CPU is rapidly hollowing out as images, texts and music lose themselves online. Senses and sensibilities are not so much recombined as isolated by the network; the individual self becomes distributed across – but no longer resides in – any one part of it. Our sensory response to the immediate environment is returned to us, but now we hear hearing, see seeing, touch touching. A new form of erotic interplay comes into being with this reflexive self. Echo's true existence begins only when Narcissus is finally lost to himself. Sound is streamed into space: cosmonauts aboard MIR have the earthly sounds of birdsong and running water piped up to them, transforming their space station into a remote orbiting lounge. Aboard the International Space Station laptops float together in a tethered cluster, forming a zero-gravity workstation.

4, 93

· 118 ·

In the autumn of 1964, as the Soviet Union sends up three cosmonauts without spacesuits in a single capsule aboard Voskhod 1, British

3, 106 architect Ron Herron draws up designs for 'an enclosed environment of colossal size that is mobile enough to traverse the world'. Visitors to the Unisphere at the New York World's Fair are already accustomed to the idea of viewing the Earth from outside space; now Herron's vision for 'A Walking City' takes the form of gigantic self-contained pods roaming about the planet on tubular telescopic legs. An early schematic can be seen in the pages of *Archigram* 5, showing a herd of them placidly wandering past the towers of a seemingly abandoned New York. Just as software is the organisation of behavioural possibilities, so too 62, 99 is the city; like Turing's Universal Machine it becomes a simulation of itself. But who directs the walking city: the city itself? Indeterminate factors such as climate and resources will guide its movements, according to Herron. As early as 1952 Claude Shannon's experiments 25, 77 at the Bell Telephone Labs reveal that a robot mouse called 'Theseus' is 75 capable of successfully running labyrinths, ringing an electric bell when it has reached its goal. There are limits to the random, which means that it can be measured and controlled. The Walking City is also a roving War Machine; the issue of *Archigram* in which it first appears bears the word 'Metropolis' in fragmented type – a location already identified as rife with sexual and social conflict.

The articulated robot arm represents the true human presence in outer space. Walking, locating and gripping machines explore alien worlds and hostile terrain: at the start of the Space Age prosthetic claws and grabs are used to handle radioactive materials in safety. In 1970, while astronauts safely operate devices for themselves on the Moon's surface, Philips presents the 'Senster' to the public at their permanent showplace in Eindhoven: a flying saucer-shaped concrete dome appropriately known as the Evoluon. Created by Edward Ihnatowicz, the Senster is a segmented autonomous arm comprising 'six independent 47 electro-hydraulic servo-systems based on the articulation of a lobster's claw'; computer-controlled and guided by sensitive microphones together with a Doppler radar system, it responds visually to the sounds made in its immediate vicinity by moving either away from or towards their source. The Evoluon's strange acoustics and the unpredictable behaviour of the crowds mean that the four-metre high Senster swivels and turns its 'head' in a seemingly intelligent manner. In 1974, two years after the last Apollo mission, the Senster is shut down and dismantled.

IBM's Deep Blue may genuinely be smart enough to beat Garry Kasparov at chess in 1997, but the supercomputer is incapable of making its own way to the New York venue let alone moving any of the pieces for itself. A number of pioneers in artificial intelligence offer a radically different approach: instead of one data-processing unit controlling limbs, movement and orientation from a central program, they suggest making the limbs themselves smart. Modelled on spiders, crabs and cockroaches, a new generation of intelligent machines starts to navigate space using limbs capable of thinking independently of each other. Crustaceans move their limbs by means of hinges, whereas vertebrates depend on pivots: so early multiple-legged prototypes, bearing warlike names like 'Genghis' and 'Attila', wander insect-like around laboratories, encountering and overcoming obstacles, learning as they move. Devoid of erotic play, in search of new information, these tiny devices do away with the human ordering of the senses. Intelligence arises out of interaction with the immediate environment. Only through the blind flailing of each limb on start-up can Genghis and Attila gain information about the domains they are about to invade. Violently intrusive, the articulated robot arm actively penetrates space, bringing it alive, making it conscious of itself. To make sense of one's surroundings leads inevitably, however, to a form of introspection: 'In the quiet of the early morning,' runs one account of the Senster's daily activities in Eindhoven, 'the machine would be found with its head down, listening to the faint noise of its own hydraulic pumps.'

10, 105

4, 97

As with the communication process, intelligence becomes illusory the moment it is simulated. Artificial intelligence is more likely to bump into human beings than dominate them. The network is all that separates flesh from the robotic arm, which represents only the smallest delay in real time. For any individual consciousness responding to a network, all events form a sequence whose orderliness eventually leads to a dreamless sleep in which figure and ground are no longer

11, 104

10, 79

distinguishable from each other – and when there is no difference between figure and ground, the result is a pattern. Genghis, Attila and the Senster are the conceptual descendants of the Walking City. Computer scientist Andrew Adamatzky creates functioning robots out of slime mould: 'This mould, or plasmodium,' he explains, 'is a naturally occurring substance with its own embedded intelligence. It propagates and searches for sources of nutrients and when it finds such sources it branches out in a series of veins of protoplasm. The plasmodium is capable of solving complex computational tasks, such as the shortest path between points and other logical calculations… By feeding it oat flakes, it grows tubes which oscillate and make it move in a certain direction carrying objects with it. We can also use light or chemical stimuli to make it grow in a certain direction.' As Marshall McLuhan and IBM ceaselessly remark throughout the Space Race, pattern recognition is information overload: the point at which extreme mobility and extreme stasis meet.

· 119 ·

Bigelow Aerospace's founder and President, Robert T. Bigelow, is a Las Vegas native who for nearly forty years has operated as a general contractor and developer in the Southwestern United States. Mr. Bigelow's primary activities have been in real estate development, as well as banking and finance. Mr. Bigelow created Bigelow Aerospace with the express purpose of revolutionizing space commerce via the development of affordable, reliable, and robust expandable space habitats.
– Bigelow Aerospace website

25, 103

'A space station is a place for relaxation for space travellers. So it is merry and bright,' observes Astro Boy, whose prosthetic form never feels the effects of deep space. The city is still the shape our future takes: supplying a protective outer shell for human exploration of the cosmos. Bigelow Aerospace specialises in the development of inflatable space stations like Genesis II, currently orbiting this planet. The use of inflatable space technology dates back to the 'Echo' communications satellite system, instrumental in first detecting the audible fallout from the universe's creation; impact-resistant materials such as Kevlar and Vectran make it possible for living units to be inflated and assembled in space where they afford their inhabitants better protection from radiation, micrometeoroids and orbital debris than conventional metal structures. Buckminster Fuller's 'tensegrity' and the inflatable outer skin suggest new relationships between humans and machines which end in the commercial colonising of space. In an open letter to NASA Administrator Charles Bolden, Robert Bigelow states that 'there are several general principles that separate "commercial space" from traditional programmes. First, "commercial" initiatives are allowed to fail.'

Space, it turns out, has always been an ideological platform: the phase from Gagarin's first orbit of the Earth to Apollo XI's landing on the Moon does not constitute a continued exploration of aerospace but the establishment of a completely new medium. Instead of Edison reciting 'Mary had a little lamb...' we have Neil Armstrong's 'One small step...' The photograph taken of the Earth from space remains the most succinct expression of this phase, and its influence can still be felt. 'You know that face on Mars?' rocket engineer Burt Rutan will later remark. 'NASA did the dumbest thing. They said it wasn't a face, it was a pile of rocks. If they'd said it was a face, they'd have full funding!' Commercial space, as Bigelow explains it, is 'less about who is doing the work than the means of procurement', with the result that the US is currently dependent upon either Russia or Elon Musk's Space Exploration Technologies Corp for its access to space. 'The Russians,' Bigelow drily observes 'are excellent capitalists, and you should expect prices to rise above the already stunning $51 million per seat that NASA is currently paying during every subsequent Soyuz contract negotiation.'

43, 96
66

Humans in space remain little more than 'spam in a can', however: dependent upon sealed environments capable of recycling their sweat, urine and breath. Gazing out of a window aboard Apollo XII, astronaut Alan Bean is suddenly aware that a single glass pane is all that separates him from death in the hard vacuum that exists beyond it. What is hollowed out must also become hardened – in other words, rendered both impermeable and crel like the 'non-human and mechanical being, constructed for an omnipresent velocity' described by Marinetti. 'When the spacesuit, space craft and space station are the architectural models,' Mark Wigley observes, 'it is understood that to leave the system is to die.' Inevitably the free-market brings with it greater issues of security, making it increasingly the personal responsibility of each individual. 'We vigorously advocate for the American airline industry as a model of safety, customer service and environmental responsibility,' runs part of the Airlines for America mission statement. Formerly the ATA, the inflexible and intrusive authority it has shown in the scanning and monitoring of passengers at airports will inevitably supply the model for the screening of those who may one day reside in commercial space.

7, 87

'Better city, better life' is the slogan for the 2010 Shanghai Expo, which seeks to provoke discussion about 'urban maladies' while a senior organiser describes the result as 'the sensation of being in a fantastic movie of light and sound or entering a theme park full of colour and attractions'. Famously visible from space, Robert Bigelow's home town of Las Vegas remains the apotheosis of Frank Lloyd Wright's modern city as 'a place for banking and prostitution and very little else' and where surveillance and security operate in plain sight. In contrast to this, there is the ground-hugging sprawl of the 'feral cities' described by Richard J Norton of the Naval War College. 'The vast size of a feral city,' he suggests, 'with its buildings, other structures, and subterranean spaces, would offer nearly perfect protection from overhead sensors, whether satellites or unmanned aerial vehicles.' Below the horizon and beneath the shadow of space resides a vast global underclass of urban squatters, economic migrants and displaced workers who find employment inside – while existing completely outside – the sealed environs of the city. It may be within their tangled labyrinths that a new form of space-age existence is discovered.

· 120 ·

Sailing towards the labyrinth was a boat containing figures which presumably represented Theseus and his companions. The design was accompanied by the words 'HIC INCLUSUS VITAM PERDIT.'
    – *W H Matthews*, Mazes and Labyrinths

DARPA is developing a species of 'small soft robots that move like jerky slithering blobs' while another of its projects, EATR, takes the form of a 'foraging robot' that gathers leaves and wood from which to generate its own electricity. In biogenetic computing, enzyme patterns are used to create new semiconductors: movement, intelligence and communication become both exploratory and more actively erotic. Humans are primarily a device for rearranging the universe; and we've grown relatively good at it over the centuries. We consciously free ourselves of our machines only by becoming one. Bowman floats weightless inside HAL 9000's mind as he seeks to dismantle it. Just as our activities occupy only a tiny electromagnetic bandwidth on Earth so, in terms of data flow, the physical plane of existence occupies a similarly small spectrum between the internal memory of the networked computer and the networked universe itself as it is slowly transformed into one vast ever-expanding computational system. Unlike rockets which consume themselves too quickly over time, spacecraft will continue to function as remote smartphones, handheld devices or applications, allowing the network to 'grow' into space for humans and robots to use. 'When you complete the primary mission, the spacecraft often survive well beyond that, so they can be re-purposed as part of a communication system,' remarks Vint Cerf, an acknowledged 'father of the Internet', regarding the prospect of an open-source interplanetary network. 'What I'm anticipating is that as we launch new missions, the previous missions' assets will become part of an interplanetary backbone.' Our furthermost extension into space must inevitably end in the separation of mind and body: a mission from which we cannot return. Similarly, to sustain the continued expansion of the human presence on Earth will involve a split and reconfiguration as total and irreversible as any of the sensory flips and inversions brought about by the introduction of the print-

77, 102

ing press, the phonograph and the network. The symbiotic relationship between humans and spaceships is reversed with the introduction of the smartphone and the tablet, which are essentially robots that cannot walk; we transport them from location to location and their interactions with the world constitute a digital spacesuit, providing us with information about our surroundings. The first 'manned landing' on Earth is going disastrously wrong, however. We have failed in establishing an exploratory relationship with our terrain. Environment is always more than just another destination. As the image of the last man, complacently hooked up to his devices, continues to proliferate on this world, we come to the realisation that we have not yet arrived on this planet. 'A pair of wings, a different respiratory system, which enabled us to travel through space, would in no way help us,' Proust reminds us, 'for if we visited Mars or Venus while keeping the same senses, they would clothe everything we could see in the same aspect as things of Earth.' A manned mission to another world that never leaves this one: the thinking behind such a venture plays upon a fundamental misreading of space, inverting it entirely. The symbolic nature of an actual human presence on Mars or elsewhere remains the one thing that our remote devices cannot replicate, favouring as they do the practical and the particular over the allegorical and the general. Nothing in the Digital Regime can exist symbolically, except when it has typographical expression. Lossless compression is possible only because most real-world data has statistical redundancy. The compression of Typographic Man may finally allow us, in Proust's words, 'to see the universe through the eyes of another, of a hundred others, to see the hundred universes that each of them sees, that each of them is'. Theseus and Ulysses can only return home by the most circuitous of routes, each becoming entwined in his own labyrinth. The Minotaur is the transformed human; part divinity, part technological prosthesis, it will always devour what remains of the human. Unlike the past, which is always with us, the future is constantly receding. Finality remains such an inexhaustible concept: it just keeps on coming, revealing just how pliable a sense of ending can be. Caught in a changeless rhythm, the countdown does not accelerate or fade in its effect: its sole purpose is to end. What comes after nine in a decimal series? Zero – which still is nothing. We do our best travelling in worlds where no welcome awaits us.

13, 96

5, 90

The Uploaded Self

A BOOK OF THE DEAD

'...even this spider and this
moonlight between the trees
and even this moment and I myself.'

Friedrich Nietzsche
The Gay Science

'Libraries will in the end become cities, said Leibniz,' observes physicist and man of letters Georg Lichtenberg in Notebook C of his *Sudelbücher* – and both today seem destined to end in ruins. In a secret labyrinth of tunnels and security zones excavated deep into the Swiss Alps, European researchers deposit a 'digital genome' designed to preserve the world's data processors together with their programming. Code written in stone lasts much longer than that which is encrypted onto discs, tapes or drives. 'In 25 years people will be astonished to see how little time must pass to render data carriers unusable because they break or because you don't have the devices anymore,' reflects one of the project partners. 'The second shock will probably be what fraction of the objects we can't use or access in 25 years and that's hard to predict.' Epic language, like epic architecture, speaks most clearly to the ages that are to come: too many surface details mislead more contemporary eyes. *Hic inclusus vitam perdit:* the words are found inscribed within a labyrinth mosaic on a Roman family tomb in the ancient necropolis at

Hadrumetum in Tunis. 'Whoever is confined here loses life.' The message and its accompanying design will later be destroyed by modern grave robbers.

The Network has become one gigantic recording device consuming everything in real time, transforming every text, every image, every sound and every experience, however trivial or unaccommodating, into a coded universe, even to the extent of replicating the background noise of its own creation. Digital scanners accidentally capture the hands of operators processing books for inclusion in online archives, their fingers protected from accumulated paper-cuts by condoms or surgical gloves. Meanwhile researchers from the European Bioinformatics Institute have 'encoded computer files totalling 739 kilobytes of hard-disk storage and with an estimated Shannon information of 5.2×106 bits into a DNA code, synthesized this DNA, sequenced it and reconstructed the original files with 100% accuracy.' Encoded into the DNA is a small cache of human documents reminiscent of those recorded onto the Voyager Golden Disc sent outside our solar system as 'a token of our sounds, our science, our images, our music, our thoughts and our feelings'. Included are all of Shakespeare's sonnets, plus a recorded excerpt from Martin Luther King's 'I Have a Dream' speech made in the early years of the Electronic Delirium, together with Crick and Watson's paper on the structure of DNA and a file describing how the data has been converted. A vast word horde and image cache has been inscribed onto dead or dying media, DNA being one of the oldest in existence. 'Meandering or labyrinthine paths, spirals, mazes actually followed in ritual (initiation) dances or symbolically represented in ritual objects,' observes Norman O Brown, 'represent the archetypal endeavours of the divine ancestor, the prototypical man, to emerge into this world, to be born.' Everything you have just read has entirely been about the future of communication design.

The European Bioinformatics Institute's researchers claim that they can scale up their DNA to store all the data in the world. In the meantime Gordon Bell argues for the power of distributed e-memory to 'change what it means to be human'. Recorded events are automatically trapped between an encroaching past and a permanently distracted present; we have come to overlook recent trends and developments as unworthy of our accelerated attention spans while at the same time organising our personal information into online equivalents of Facebook's 'foundational narrative timeline'. The thought that the detail of our lives can be preserved for our uploaded selves to experience sometime in the future leads to the prospect of the data processor itself emerging as a new behavioural entity. Illuminated by the bright digital blast of its existence, the self is finally wiped out in order to become an immortal consciousness. It is no longer the oblivion threatened by a thermonuclear blast but the one which is triggered by the digital preservation of the self as some form of utopian upload that alarms us most today. Consciousness becomes a flat silhouette projected onto a ruined wall somewhere in Hiroshima: the shadow cast by programming code upon the world. Remembering all things all of the time, the digital self will be resurrected only as memory. The prospect that this resurrection may somehow take place within a computing environment suggests that a form of immortality is possible but only at the risk of bringing to an end our hitherto dynamic relationship with technology.

What appears in retrospect to have been a logical progression is in actuality a state of violent unrest. Machines and humans have continually penetrated each other over the centuries, producing between them an unresolved cultural flux in which nothing remains fixed. Technology is, in other words, best understood as the living depiction of humanity's struggle with itself. Media theory and machine history have so far allowed us snapshots of a process that is constantly taking place both within us and outside us at the same time. If we look at the history of the Network itself rather than the different platforms and applications that have developed through it, we find that we have been witnessing the expansion of one single networked existence. Far from being inert, it is as alive as any multi-cellular organism would be, expanding and contracting in response to external stimuli, its actions and reactions triggered by consumers clicking their way through the system. Such is the rapidly evolving nature of this strange new creature that any account of the Network's condition must inevitably be out of date before it appears. As mere stages in the Network's restless development, the World Wide Web and the Internet are consequently historical moments that are without moment. Totally consumed by its digital environment, the Uploaded Self will hear and see everything except what is actually happening to it.

In the end we are trading the consolations of one dangerous landscape for another. In their shift between mineral and biological levels of existence, both represent useful distractions. The Uploaded Self offers the bodiless sensation of memory, invading present time from anywhere within the known universe. Correspondingly, the challenge posed to humanity by the physical vastness of space is forcing us to renounce the possibility of being able to experience an increasingly hostile cosmos with our bodily senses. Almost half a century after the first Apollo Moon landings we remain physiologically as earthbound as ever. The creeping awareness that we, in contradiction to everything that the Electronic Delirium has promised us, are 'here to stay' is deeply unsettling. It imposes itself upon our values and perceptions, shaping their development in the same way that the Death of God helped form some of the deeper folds and creases in twentieth-century thought. Requiring nothing to disprove them, fantasies eventually become fixed and unchanging. A mathematical study of black holes conducted by physicist Leonard Susskind has led to a theory that the three-dimensional universe is in reality a holographic projection of two-dimensional data stored along the boundary of that universe. According to this particular scenario, we have already passed through the Event Horizon, leaving behind only that information which distinguishes each of us as a unique being. Given a computer powerful enough, this information could then be processed to create an exact duplicate of our existence in a similarly duplicated projection of physical space. The concept has, Susskind believes, 'already reached a consensus' among theoretical physicists. From first to last, the Uploaded Self maintains itself as an article of faith.

On 8 November 2012 Larry Burns, United States District Court Judge for the Southern District of California, sentences Jared Loughner to seven life sentences without the possibility of parole plus 140 years in prison for his part in the Tucson supermarket shooting of January 2011. Six of the life sentences, Judge Burns explains, represent the six slain victims while the seventh is for the attempted assassination of United States Representative Gabrielle Giffords. 'Each of those victims was important,' he continues. 'It reflects each of those individual lives. There is a symbolic nature in this.' Loughner has previously pleaded guilty to nineteen counts of murder or attempted murder and has been ruled mentally competent to stand trial. Although partially paralyzed by the shooting and able to speak only with difficulty, Gabrielle Giffords attends the sentencing in the company of her husband, retired astronaut Mark Kelly. Through him, Giffords tells Loughner that she is now 'done thinking about' her would-be killer. 'Consciousness,' Schrödinger argues, 'is never experienced in the plural, only in the singular.' And so too is death. The person who remembers something is not always the one who experiences it – even when they share the same name and identity. The first and most important function of a Book of the Dead is therefore to remind the recently departed that they have in fact died.

Something written in this world for the next, a Book of the Dead sets every act, thought and gesture against eternity to be experienced once more. Sex and death are fleeting interruptions to an eternal recurrence by which, according to Nietzsche's formula, 'every pain and every joy and every thought and sigh and everything unspeakably small or great in your life must return to you, all in the same succession and sequence'. Programming a machine to be conscious of itself may also require that it become aware of its own end: self-awareness being also a singular awareness of your own death. A digitally replicated universe is one that spontaneously generates and projects its own designs to the extent that you can choose not only to experience but also to *become* 'this spider and this moonlight between the trees'. Sooner or later all labyrinths double back on themselves. Design is finally sensation made aware of itself. In this respect it is both precise and deliberate. 'If, as Leibniz has prophesied, libraries one day become cities,' Lichtenberg continues in Notebook J, some twenty years later, 'there will still be dark and dismal streets and alleyways as there are now.' And some of them may still connect us up to this haunted planet deep in the loneliness of space.

Notes and References

BY CHAPTER AND SECTION

How to Influence the Past: A Book for the Living

Plutarch, 'Sertorius', *Makers of Rome,* Translated with an Introduction by Ian Scott-Kilvert, Penguin, Harmondsworth, 1965, p. 195

Francis Bacon, *Of the Proficience and Advancement of Learning, Divine and Human,* Project Gutenberg HTML, *http://www.gutenberg.org/dirs/etext04/adlr10h.htm,* accessed 4.5.13

Welcome to the Labyrinth

2.

Mark Wigley, 'The Architectural Brain' in *Network Practices*: New Strategies in Architecture and Design, edited by Anthony Burke and Therese Tierney, Princeton Architectural Press, New York, 2007, p. 36

Warren Chalk, 'Hardware of a New World', *Architectural Forum,* October 1966, quoted in Hadas A Steiner, *Beyond Archigram*: The Structure of Circulation, Routledge, New York and Abingdon, 2009, p. 199

Walter Benjamin, *The Work of Art in the Age of Mechanical Reproduction*, 'But the instant the criterion of authenticity ceases to be applicable to artistic production, the total function of art is reversed. Instead of being based on ritual, it begins to be based on another practice – politics.' *http://www.marxists.org/reference/subject/philosophy/works/ge/benjamin.htm,* accessed 13.7.10

3.

'China's Dash for Freedom', *The Economist*, Volume 388, Number 8591, 2-8 August 2008, p.11

Nick Tosches, 'Dubai's The Limit', *Vanity Fair*, Number 550, June 2006, p. 130

Johan Hari, 'The Dark Side of Dubai', *Independent*, 6, April 2009, http://www.independent.co.uk/voices/commentators/johann-hari/the-dark-side-of-dubai-1664368.html, accessed 13.7.10

Richard Wagner on *Parsifal* stage directions, http://www.musicwithease.com/parsifal-wagner-plot.html, accessed 15.3.13

4.

Brion Gysin, *The Process*, Granada Publishing, St Albans, 1973, p. 218

Mark Wigley, 'The Architectural Brain' in *Network Practices*: New Strategies in Architecture and Design, edited by Anthony Burke and Therese Tierney, Princeton Architectural Press, New York, 2007, pp. 30-31

5.

Ovid, Book VIII: 140-174, *Metamorphoses*, translated with an introduction by Mary M. Innes, Penguin Books, London, 1955, p. 183

Borges, Jorge Luis, 'Death and the Compass', translated by Anthony Kerrigan, from *Fictions*, Calder and Boyars 1974, London, p. 128

Homer, Book 18: 590-617, *The Iliad*, translated with an introduction by Martin Hammond, London, Penguin Books, 1987, p. 310

7.

Jerry Bruckheimer's comment passed on to the author in conversation with Trevor Horn, 13.1.09

Augé, Marc, *Non-places*: Introduction to an Anthropology of Supermodernity, translated by John Howe, Verso, London, 1995

'Corporate Nomads', *Time*, September 29, 1967

George Nelson, 'Architecture for the New Itinerants' *Saturday Review* 22 April 1967, quoted Alastair Gordon, *Naked Airport*: A Cultural History of the World's Most Revolutionary Structure, University of Chicago Press, Chicago and London, 2008, p.214

Archigram 8 (1968) un-paginated reference

Unidentified pilot, *Conquest of the Skies*, quoted in Gordon 2008, p. 167

Rogers Stirk Harbour and Partners website, http://www.rsh-p.com/work/
buildings/terminal_5_heathrow_airport/completed, accessed 2.7.09

8.

Alastair Gordon, *Naked Airport*: A Cultural History of the World's Most
Revolutionary Structure, University of Chicago Press, Chicago and London,
2008, p. 238

Marshall McLuhan, *Understanding Me*: Lectures and Interviews, edited by
Stephanie McLuhan and David Staines, The MIT Press, Massachusetts, 2003,
p. 120

'Watching Nervously', *The Economist*, 2-8 May 2009, p. 71

'Getting Inside Your Head, *The Sunday Times*, 3 May 2009, p. 13

9.

David Greene, 'Instant City Children's Primer', *Architectural Design*, May 1969,
p. 276,

Marc Andresseen, quoted in Kurt Eichenwald, 'Facebook's Future', *Vanity Fair*
633, May 2013, p. 164

Jeffrey Huang, Notes from a lecture given December 2007, http://www.
lunchoverip.com/2007/12/a-designer-at-t.html, accessed 10.10.08

Songdo Website, Gale International Real Estate Development http://www.
songdo.com/home/why-songdo/a-brand-new-city.aspx, accessed 12.7.09,
emphasis in the original post

10.

Jonathan Fildes, 'Barcode replacement shown off', BBC News, July 27, 2009
http://news.bbc.co.uk/1/hi/technology/8170027.stm, accessed 26.8.09

Camera Culture, Computation Camera and Photography, Fall Course 2009,
http://cameraculture.media.mit.edu/, accessed 26.8.09

11.

Jean-Luc Godard, *Alphaville*, translated by Peter Whitehead, Faber and Faber
Ltd, London, 2000, p. 39

Josh McHugh, 'Attention, Shoppers: You Can Now Speed Straight Thru Checkout
Lines!', *Wired* 12.7, July 2004, p. 152

'Tracking trash: Project aims to raise awareness of how garbage impacts the environment' July 15, 2009, http://web.mit.edu/newsoffice/2009/trash-0715. html, accessed 26.8.09

Downer, Green and Carrillo; Clarke, Rapuano and Holleran; Harrison and Abromowitz, *The New York Municipal Airport at Idlewild*, (New York un-paginated, 1946) quoted Alastair Gordon, *Naked Airport*: A Cultural History of the World's Most Revolutionary Structure, University of Chicago Press, Chicago and London, 2008, p. 154

Neha, Dutt, 'RFID-enabled checkpoint eases your shopping', the Design Blog, August 6, 2009 http://www.thedesignblog.org/entry/rfid-enabled-checkpoint-eases-your-shopping/, accessed 5.9.09

12.

William Shakespeare, *A Midsummer Night's Dream*, Act III, Scene 1, The Arden Shakespeare edition, edited by Harold F. Brooks, Methuen, London and New York, 1986, p. 55

Mike Davis, 'War-Mart "Revolution" in warfare slouches toward Baghdad: It's all in the network' *San Francisco Chronicle*, March 9, 2003, http://www. sfgate.com/cgi-bin/article.cgi?file=/chronicle/archive/2003/03/09/IN8529. DTL#ixzzoRG4wTTwo, accessed 15.9.09

Kevin Kelly, Internet Mapping Project http://www.kk.org/internet-mapping/, accessed 16.9.09

Mara Vanina Osés, the Internet Mapping Project, PDF download http:// psiytecnologia.files.wordpress.com/2009/06/the-internet-mapping-project2. pdf, accessed 6.9.09. Note: spelling, phrasing and grammar retained from Professor Osés's original text.

Olaf Arndt, 'Labyrinth and Camp: On the presence of a primal image in times of crisis', *El Dorado*: On the Promise of Human Rights, Kerber Verlag, Bielefeld, 2009, p. 25

13.

Mike Davis, 'War-Mart "Revolution" in warfare slouches toward Baghdad: It's all in the network' *San Francisco Chronicle*, March 9, 2003 http://www. sfgate.com/cgi-bin/article.cgi?file=/chronicle/archive/2003/03/09/IN8529. DTL#ixzzoRG4wTTwo, accessed 15.9.09

Hitachi hard disk advertising copy, taken from a series of adverts in *Wired* magazine appearing between November 2004 (issue 12/11) and January 2005 (issue 13/01)

Akira Suzuki, *Do Android Crows Fly Over the Skies of an Electronic Tokyo?* The Interactive Urban Landscape of Japan, Architectural Association, London, 2001, pp. 51-52

The Future is What Happens After You're Dead
14.
Gary Wolf, 'Channelling McLuhan', *Wired* 2.01, January 1996, p. 51

15.
Herbert Marcuse, *Eros and Civilization*: A Philosophical Inquiry into Freud, Routledge, Abingdon, 2006, p. 3

Marshall McLuhan, 'Towards an Inclusive Consciousness', *Understanding Me*: Lectures and Interviews, Stephanie McLuhan and David Staines (eds.), MIT Press, Massachusetts, 2003, p. 125

H G Wells, *Mind at the End of Its Tether*, reprinted in *The Last Books of H G Wells*: The Happy Turning & Mind at the End of Its Tether, e-book, Provenance Editions, New York, 2006

16.
Bullwinkle J Moose, quoted in *The Whole Pop Catalogue*: the Berkeley Pop Culture Project, Avon Books, New York, 1992, p. 463

Andew Tilin, 'You Are About to Crash', *Wired* 10.04, April 2002, p. 97

Jerry Garrett 'Fifty, Finned and Fabulous', *New York Times*, 20 May 2007 http://www.nytimes.com/2007/05/20/automobiles/collectibles/20FIFTY.html, accessed 10.10.09

17.
The Sunday Times, 3 May 2009, front cover copy.

Matthew Moore, '50 most annoying things about the internet', *Daily Telegraph* post, published 7:00AM BST 16 October 200 http://www.telegraph.co.uk/technology/news/6338303/50-most-annoying-things-about-the-internet.html, accessed 19.10.09

18.

Kim Stanley Robinson, in conversation with the author, 1.12.05

19.

Marshall McLuhan, 'TV News as a New Mythic Form', television interview with
   Tom Wolfe, 1970, reproduced in *Understanding Me*: Lectures and Interviews,
   Stephanie McLuhan and David Staines (eds.), MIT Press, Massachusetts, 2003,
   p. 172
Friedrich A Kittler, *Gramophone, Film, Typewriter*, translated with an introduction
   by Geoffrey Winthrop-Young and Michael Wutz, Stanford University Press,
   California, 1999, p. xxxix
Marshall McLuhan, 'Towards an Inclusive Consciousness', *Understanding Me*:
   Lectures and Interviews, Stephanie McLuhan and David Staines (eds.), MIT
   Press, Massachusetts, 2003, p. 129
Eric Norden, 'The Playboy Interview: Marshall McLuhan', *Playboy* Magazine,
   March 1969, http://www.nextnature.net/?p=1025, accessed 28.10.09
Marshall McLuhan, 'Cybernetics and Human Culture', *Understanding Me*:
   Lectures and Interviews, Stephanie McLuhan and David Staines (eds.), MIT
   Press 2003, Massachusetts p. 49

20.

'Space Probe!', *Archigram* 4, 1964, p. 1
Marshall McLuhan, 'Cybernetics and Human Culture', *Understanding Me*:
   Lectures and Interviews, Stephanie McLuhan and David Staines (eds.), MIT
   Press 2003, Massachusetts, p. 49
Herbert Marcuse, *One-Dimensional Man*: Studies in the Ideology of Advanced
   Industrial Society, Routledge, Abingdon and New York, 2007, p. 163
Ralph Waldo Emerson, 'Works and Days', reprinted in *Of Men and Machines*,
   edited by Arthur O Lewis Jr, Dutton Paperbacks, New York, 1963, p. 64
Norbert Wiener, *Human Use of Human Beings*, quoted in *Dark Hero of the
   Information Age*: In Search of Norbert Wiener, Flo Conway and Jim
   Siegelman, Basic Books, New York, 2005, p. 277

21.

Herbert Marcuse, *One-Dimensional Man*: Studies in the Ideology of Advanced
   Industrial Society, Routledge, Abingdon and New York, 2007, p. xlvii

Herbert Marcuse, *Eros and Civilization*: A Philosophical Inquiry into Freud, Routledge, Abingdon, 2006, p. 41 (citations within quote from Sigmund Freud's *An Outline of Psychoanalysis*, p. 26)

Marshall McLuhan, letter to Bill Jovanovich, quoted in *Marshall McLuhan*: Escape Into Understanding, W Terence Gordon, Basic Books, New York, 1997, p. 218

Herbert Marcuse, *One-Dimensional Man*: Studies in the Ideology of Advanced Industrial Society, Routledge, Abingdon and New York, 2007, p. 193

22.

Marshall McLuhan, 'Technology, the Media and Culture', *Understanding Me*: Lectures and Interviews, Stephanie McLuhan and David Staines (eds.), MIT Press, Massachusetts, 2003, p. 23

Marshall McLuhan: Escape into Understanding, W Terence Gordon, Basic Books, New York, 1997, p. 246

Marshall McLuhan, 'Technology, the Media and Culture', *Understanding Me*: Lectures and Interviews, Stephanie McLuhan and David Staines (eds.), MIT Press, Massachusetts, 2003, pp. 23-4

Konrad Wachsmann, *The Turning Point of Building*, quoted in *Network Practices*: New Strategies in Architecture and Design, edited by Anthony Burke and Therese Tierney, Princeton Architectural Press, New York, 2007, p. 44

Herbert Marcuse, *One-Dimensional Man*: Studies in the Ideology of Advanced Industrial Society, Routledge, Abingdon and New York, 2007, p. 231

Gilles Deleuze & Félix Guattari, *Anti-Oedipus*: Capitalism and Schizophrenia, translated by Robert Hurley, Mark Seem and Helen R. Lane, Viking Press, New York, 1977, pp. 7 and 348

23.

Marshall McLuhan, *Understanding Media*: The Extensions of Man, Routledge, London, 2002, p. 366

Stanley Milgram, 'The Perils of Obedience' *Harper's Magazine*, December 1973, p. 76, abridged and adapted from 'Obedience and Authority' http://www.harpers.org/archive/1973/12/page/0078, accessed 12.11.09

Herbert Marcuse, *One-Dimensional Man*: Studies in the Ideology of Advanced Industrial Society, Routledge, Abingdon and New York, 2007, pp. 11 and 53

Eric Norden, 'The Playboy Interview: Marshall McLuhan', *Playboy* Magazine, March 1969, http://www.nextnature.net/?p=1025, accessed 28.10.09

24.

Herbert Marcuse, *One-Dimensional Man*: Studies in the Ideology of Advanced Industrial Society, Routledge, Abingdon and New York, 2007, p. 63

Marshall McLuhan, 'Technology, the Media and Culture', *Understanding Me*: Lectures and Interviews, Stephanie McLuhan and David Staines (eds.), MIT Press, Massachusetts, 2003, p. 25

Herbert Marcuse, *One-Dimensional Man*: Studies in the Ideology of Advanced Industrial Society, Routledge, Abingdon and New York, 2007, pp.198 and 213

Marshall McLuhan, *Understanding Media*: The Extensions of Man, Routledge, London, 2002, p. 246

Marshall McLuhan and David Carson, *The Book of Probes*, Ginko Press, California, 2003, pp. 30-33

Herbert Marcuse, *One-Dimensional Man*: Studies in the Ideology of Advanced Industrial Society, Routledge, Abingdon and New York, 2007, p. 70

25.

Marshall McLuhan, *Understanding Media*: The Extensions of Man, Routledge, London, 2002, p. 183

Michael Wolff, 'Rupert Murdoch to Internet: It's War!', *Vanity Fair* 591, November 2009, p. 110

Herbert Marcuse, *One-Dimensional Man*: Studies in the Ideology of Advanced Industrial Society, Routledge, Abingdon and New York, 2007, p. 250

Gary Wolf, 'Channelling McLuhan', *Wired* 2.01 January 1996, p. 51

Mika Taanila, *The Future is Not What It Used to Be*, Kinotar films, 2003

Towards A Sexual History of Machines
26.

Robert Linhart, *The Assembly Line*, translated by Margaret Crosland, John Calder, London, 1981, pp. 13-14

27.

Georges Bataille, 'The Solar Anus', *Visions of Excess*: Selected Writings 1927-1939, translated by Allan Stoekel with Carl R Lovett and Donald M Leslie, Jr, University of Minnesota Press, Minneapolis, 1985, p. 6

Charles R Walker, *Toward the Automatic Factory*, Yale University Press, 1957, p. 104, quoted in Marcuse, *One-Dimensional Man*, p. 29

Jean-Paul Sartre, *Critique de la mison dialectique*, Gallimard 1960, p.290, quoted in Marcuse, *One-Dimensional Man*, p. 29

28.

Gilbert Simondon, *On the Mode of Existence of Technical Objects*, translated by Ninian Mellamphy with a Preface by John Hart, University of Western Ontario, June 1980, online PDF, pp. 19-20 http://nsrnicek.googlepages.com/ SimondonGilbert.OnTheModeOfExistence.pdf, accessed 5.12.09

Georges Bataille, *Eroticism*, translated by Mary Dalwood, Marion Boyars, London, 1987, p. 29

Sigmund Freud, *Civilization and Its Discontents*, SE, vol. 21, Hogarth Press, London, 1975, pp. 66-67

Herbert Marcuse, *Eros and Civilization*: A Philosophical Inquiry into Freud', Abingdon, Routledge, 2006, p. 86

29.

Sigmund Freud, *Civilization and Its Discontents*, SE, vol. 21, Hogarth Press, London, 1975, p. 91

Herbert Marcuse, *One-Dimensional Man*: Studies in the Ideology of Advanced Industrial Society, Routledge, Abingdon and New York, 2007, p. 6

30.

Arthur Rimbaud, Letter to Paul Demeny, May 15, 1871, http://www.mag4.net/ Rimbaud/DocumentsE1.html, bilingual site, accessed 6.12.09

Jerrold Northrop Moore, *Sound Revolutions*: A Biography of Fred Gaisberg, Founding Father of Commercial Sound Recording, Sanctuary Music Library, Sanctuary Publishing, London 1999, p. 67

Friedrich A Kittler, *Gramophone, Film, Typewriter*, translated with an Introduction by Geoffrey Winthrop-Young and Micheal Wutz, Stanford University Press, California, 1999

31.

One of One: Snapshots in Sound, CD release, Dish Records, Californai, dish 002, no publication year. The phrases in speech marks at the start of each paragraph in this section are transcribed directly from 'Sal Boo', track 16, *One of One: Snapshots in Sound*

Friedrich Nietzsche, *Human, All Too Human*, translated by Marion Faber and
    Stephen Lehmann, Penguin Books, London, 1994, p. 117

32.
Ken Jacobs, *Film-Makers Cooperative Catalogue* No 5, p. 167, quoted in
    P Adams Sitney, *Visionary Film*: The American Avant-Garde 1943-1978, second
    edition, Oxford University Press, New York, 1979, p. 368

33.
Norman O Brown, *Love's Body*, Random House, London, 1966, p. 99
Marshall McLuhan, *Understanding Media*: The Extensions of Man, Routledge,
    London, 2001, pp. 8-9
Jonathan Gathorne-Hardy, *Alfred C Kinsey*: Sex The Measure of All Things,
    Pimlico, London, 1999, p. 307
Herbert Marcuse, *One-Dimensional Man*: Studies in the Ideology of Advanced
    Industrial Society, Routledge, Sanctuary Music Library, 2002, p. 189
Stephen Fried, 'The New Sexperts', *Vanity Fair*, volume 55, number 12, December
    1992, p. 58

34.
Gene Swenson, 'What Is Pop Art? Interviews with eight painters' *Art News*,
    November 1963, Warhol interview reproduced in *Pop Art Redefined*, edited by
    John Russell and Suzi Gablik, Thames and Hudson, London, 1969, p. 118
Oliver Wendell Holmes, quoted in 'Who Needs It?' *The Economist*, 1-7 May 2010,
    p. 12
Lewis Mumford, 'The Monastery and the Clock', reprinted in *Of Men and
    Machines*, edited by Arthur O Lewis Jr, Dutton Paperbacks, New York, 1963,
    p. 63

35.
Walter Benjamin, 'Paris, Capital of the Nineteenth Century', quoted in *Nietzsche
    and 'An Architecture of Our Minds'*, edited by Alexandre Kostka and Irving
    Wohlfarth, Getty Research Institute for the History of Art and the Humanities,
    Los Angeles, 1999 p. 160

36.

Alfred Jarry, *The Supermale*, translated by Barbara Wright, Jonathan Cape, London, 1968, p. 40

H G Wells, *New York Times*, April 17, 1927, http://erkelzaar.tsudao.com/reviews/ H.G.Wells_on_Metropolis%201927.htm, accessed 23.8.10

37.

C G Gens-d'Armes, quoted in Roger Shattuck, *The Banquet Years*, The Origins of the Avant-Garde in France 1885 to World War I, Vintage revised edition, New York, 1968, p. 192

Friedrich Nietzsche, *The Birth of Tragedy* Out of the Spirit of Music, translated by Shaun Whiteside, edited by Michael Tanner, Penguin Books, London, 1993, p. 46

Friedrich Nietzsche, *Untimely Meditations*, II 5, translated by R J Hollingdale, Cambridge University Press, Cambridge, 1997, p. 85

38.

Marshall McLuhan, 'The Playboy Interview', March 1969, copyright 1994 by *Playboy*, http://www.mcluhanmedia.com/mmclpb01.html, accessed 13.4.08

Tim Barribeau, 'A Device That Lets You Type With Your Mind', io9 website, http:// io9.com/5423338/a-device-that-lets-you-type-with-your-mind, accessed 12.12.09

Joel Johnson, Kuka: Robot Ascetic Transcribes Bible, Boing Boing Gadgets, http://gadgets.boingboing.net/2007/10/25/kuka-robot-ascetic-i.html, accessed 13.12.09

The Dream of Venus
39.

Garrett Putman Serviss, *Edison's Conquest of Mars*, http://www.gutenberg.org/ files/19141/19141-h/19141-h.htm, accessed 13.4.10

Jules Verne, *Twenty Thousand Leagues Under the Sea*, quoted in *The Artificial Kingdom*, A Treasury of the Kitsch Experience, Celeste Olalquiaga, Bloomsbury, London, 1999, p. 104

40.

Norman O Brown, *Life Against Death*: The Psychoanalytical Meaning of History,

Second Edition, Wesleyan University Press, Connecticut, 1985, p. 12
M Horkheimer and T W Adorno, *Dialektic der Aufklarung*, translated by
    H Marcuse, quoted in *One-Dimensional Man*, p. 161
Norman O Brown, *Love's Body*, Random House, New York, 1966, p. 48

41.

Walter Benjamin, *The Arcades Project*, translated by Howard Liland and Kevin
    McKaughlin, The Belknap Press of Harvard University Press, Cambridge,
    Massachusetts and London, England, 1999, p. 176
H G Wells, *The War of the Worlds*, Penguin Classics, London, 2005, p. 166
Christopher McIntosh, *The Swan King*: Ludwig II of Bavaria, Tauris Parker
    Paperbacks, London and New York, 2009, p. 139

42.

All direct quotes in this section are taken from Rem Koolhaas, *Delirious New York*:
A Retroactive Manifesto for Manhattan, The Monacelli Press, New York, 1994
except for the description of the Hippodrome's mechanical stage, which is taken
from Irving Kolodin's 'Speaking of Programmes', quoted in Fernand Ouellette,
Edgard Varèse: A Musical Biography, Grossman Publishers, New York, 1968,
p. 48, and the concluding quotation from Walter Benjamin, *The Arcades Project*,
translated by Howard Liland and Kevin McKaughlin, The Belknap Press of Harvard
University Press, Cambridge, Massachusetts and London, England, 1999, p. 843.

44.

Salvador Dali, 'Announcement', 'written euphonically by Dali to be read aloud
    upon his first visit to New York' in 1936, quoted with a descriptive comment
    from Julien Levy in Ian Gibson, *The Shameful Life of Salvador Dali*, Faber and
    Faber Ltd, London, 1997, p. 368
Theodor Adorno, quoted in Mike Davis, *City of Quartz*: Excavating the Future in
    Los Angeles, New Edition, Verso, London and New York, 2006, p. 48

46.

All direct quotes in this and the following section are taken from Ingrid Schaffner,
    Dali's Dream of Venus: The Surrealist Funhouse From the 1939 World's Fair,
    Princeton Architectural Press, New York, 2002, except where otherwise
    indicated.

Norman O Brown, *Love's Body*, Random House, New York, 1966, pp. 81-82.

'World's Fairs. Pay As You Enter', *Time* Magazine, Chicago, 16 June 1939, quoted in Ian Gibson, *The Shameful Life of Salvador Dali*, Faber and Faber Ltd, London, 1997, p. 391

Du Pont press release, quoted in Susannah Handley, *Nylon*: The Manmade Fashion Revolution, A Celebration of Design from Art Silk to Nylon and Thinking Fibres, Bloomsbury, London, 1999, p. 43

47.

Copy on back of photograph featuring Evonne Kummer and 'Blitzcreak' issued by Director of Publicity, New York World's Fair, New York Public Library Digital Gallery, http://digitalgallery.nypl.org/nypldigital/dgkeysearchdetail. cfm?strucID=1791645&imageID=1652643#_seemore, accessed 19.8.10

Salvador Dali, 'L'Ane Pourri', Le Surréalisme au service de la revolution, 1930, quoted in Walter Benjamin, *The Arcades Project*, translated by Howard Liland and Kevin McKaughlin, The Belknap Press of Harvard University Press, Cambridge, Massachusetts and London, England, 1999, p. 547

48.

Official Guide Book, New York's World's Fair, quoted in Rem Koolhaas, *Delirious New York*: A Retroactive Manifesto for Manhattan, The Monacelli Press, 1994, p. 276

Christopher McIntosh, *The Swan King*: Ludwig II of Bavaria, Tauris Parker Paperbacks, London and New York, 2009, p. 135

Vannevar Bush, 'As We May Think', *The Atlantic*, July 1945, available at http://www. theatlantic.com/magazine/archive/1969/12/as-we-may-think/3881/, accessed 6.9.10

Herbert Marcuse, *Eros and Civilization*: A Philosophical Inquiry into Freud', Routledge, Abingdon, 2006, p. 119

Richard S Shaver, quoted at length in the chapter 'Remembering Lemuria' from Jim Schnabel, *Dark White*: Aliens, Abductions and the UFO Obsession, Hamish Hamilton, London, 1994

49.

Sigmund Freud, *New Introductory Lectures* pp. 38-9 (1949), quoted in Norman O Brown, *Love's Body*, Random House, 1966, p. 39

50.
Walter Benjamin, *The Arcades Project*, translated by Howard Liland and Kevin
  McKaughlin, The Belknap Press of Harvard University Press, Cambridge,
  Massachusetts and London, 1999, p. 551

Duration, Or the Birth of Chance Out of the Spirit of Music
51.
Thomas Mann, *Doctor Faustus*, translated by John E Woods, Vintage International
  Edition, New York, 1997, p. 68
Walter Benjamin, *The Arcades Project*, translated by Howard Liland and Kevin
  McKaughlin, The Belknap Press of Harvard University Press, Cambridge,
  Massachusetts and London, 1999, p. 551
Ornella Volta, *Satie Seen Through his Letters*, translated by Michael Bullock,
  Marion Boyars, London and New York, 1989, p. 175

52.
All direct quotes in this section are taken from Fernand Ouellette, *Edgard Varèse*:
A Musical Biography, Grossman Publishers, New York, 1968, except the copy for
Philips advertising copy, which is reproduced from the sleeve notes to *Popular
Electronics*: Early Dutch Electronic Music from Philips Research Laboratories 1956-
1963, CD release, Basta 30.9141.2

53.
Edgard Varèse, quoted in Louise Varèse, *Varèse: A Looking Glass Diary,Volume I*:
  1883-1928, Davis-Poynter, London, 1973, p. 105
Francesco Cangiullo, 'Detonation: Synthesis of All Theatre', reproduced in
  Michael Kirby, *Futurist Performance*, E P Dutton & Co Inc, New York, 1971,
  p. 247
Fernand Ouellette, *Edgard Varèse*: A Musical Biography, Grossman Publishers,
  New York, 1968, p. 106
Karel Čapek, R U R (*Rossum's Universal Robots*), reprinted in *Of Men and
  Machines*, edited by Arthur O Lewis Jr, Dutton Paperbacks, New York, 1963,
  p. 36

54.

Vannevar Bush, 'As We May Think', *The Atlantic*, July 1945, available at http://www.theatlantic.com/magazine/archive/1969/12/as-we-may-think/3881/, accessed 6.9.10

Varèse's telegram to Cage quoted in *Begin Again*: A Biography of John Cage, Kenneth Silverman, Alfred A Knopf, New York, 2010, p.43

55.

Laurence Sterne, *The Life and Opinions of Tristram Shandy, Gentleman*, Penguin Books, London, 2003, p. 467

John Cage, quoted in Calvin Tomkins, *Ahead of the Game*: Four Versions of Avant-garde, Penguin Books, Harmondsworth, 1968, p. 100

Lewis Mumford, 'The Monastery and the Clock', reprinted in *Of Men and Machines*, edited by Arthur O Lewis Jr, Dutton Paperbacks, New York, 1963, p. 62

Edie: American Girl, Jean Stein, edited with George Plimpton, Grove Press 1982, p. 235

56.

Merce Cunningham in conversation with Jacqueline Lesschaeve, *The Dancer and the Dance*, Marion Boyars, revised edition 1990, New York and London, pp. 140-141

John Cage, '45' For A Speaker', *Silence*: Lectures and Writings, Wesleyan University Press, Connecticut, 1961, p. 174

Ibid, p. x

John Cage with Daniel Charles, *For the Birds*, Marion Boyars, New York and London, 1981, p. 89

Marshall McLuhan and David Carson, *The Book of Probes*, Ginko Press, California, 2003, p. 294

57.

Robert Watts, 'Casual Event', first published in *Events*, Fluxus eds, New York, 1963, quoted in Sally Banes, *Greenwich Village 1963*: Avant-Garde Performance and the Effervescent Body, Duke University Press, Durham and London, 1993, p. 116

Norman O Brown, *Life against Death*: The Psychoanalytical Meaning of History, Second Edition, Wesleyan University Press, Connecticut, 1985 p. 297

John Cage, 'Lecture on Nothing', *Silence*: Lectures and Writings, Wesleyan
    University Press, Connecticut, 1961, p. 111
All quotes from John Cage in this section are taken from Kenneth Silverman,
    Begin Again: A Biography of John Cage, Alfred A Knopf, New York, 2010

58.
Laurence Sterne, *The Life and Opinions of Tristram Shandy, Gentleman*, Penguin
    Books, London, 2003, p. 335
John Cage, 'Where Are We Going? And What Are We Doing?', *Silence*: Lectures
    and Writings, Wesleyan University Press, Connecticut, 1961, p. 222.
Merce Cunningham, in conversation with the author, 31.7.06
George Maciunas, quoted in *Joseph Beuys*, Caroline Tisdall, Thames and
    Hudson, London, 1979, p. 84
Karel Čapek, *R U R (Rossum's Universal Robots)*, reprinted in *Of Men and
    Machines*, edited by Arthur O Lewis Jr., Dutton Paperbacks, New York, 1963,
    p. 19

59.
Mao Tsetung, quoted with this spelling of Mao's name in the frontispiece to *Taking
    Tiger Mountain by Strategy*, A Modern Revolutionary Peking Opera, Foreign
    Languages Press, Peking, 1971
Norman O Brown, *Life against Death*: The Psychoanalytical Meaning of History,
    Second Edition, Wesleyan University Press, Connecticut 1985 p. 233
Jill Johnson, *Jasper Johns*: Privileged Information, Thames and Hudson, London,
    1996, pp. 118-119
Herbert Marcuse, quoting Ernest Schachtel, *Eros and Civilization*: A Philosophical
    Inquiry into Freud, Routledge, Abington, 2006, p. 39
Thomas Edison, 'The Perfected Phonograph' (1888), quoted in Greg Milner,
    Perfecting Sound Forever, The Story of Recorded Music, Granta, London, 2009,
    p. 36
Herbert Marcuse, *Eros and Civilization*: A Philosophical Inquiry into Freud,
    Routledge, Abington, 2006, p. 145

60.
Laurence Sterne, *The Life and Opinions of Tristram Shandy, Gentleman*, Penguin
    Books, London, 2003, p. 204

Description of Cage's blank pages, Kenneth Silverman, *Begin Again*: A Biography
of John Cage, Alfred A Knopf, New York, 2010, p. 96
John Cage with Daniel Charles, *For the Birds*, Marion Boyars, New York and
London, 1981, p. 117
Laurence Sterne, *The Life and Opinions of Tristram Shandy, Gentleman*, Penguin
Books, London, 2003, p. 431
John Cage with Daniel Charles, *For the Birds*, p. 114
Marshall McLuhan, *The Gutenberg Galaxy*: The Making of Typographic Man,
Toronto University Press, Toronto, 1962 p. 45
Norman O Brown, *Life against Death*: The Psychoanalytical Meaning of History,
Second Edition, Wesleyan University Press, Connecticut, 1985, p. 69 and
p. 292

61.
Thomas Mann, *Doctor Faustus*, translated by John E Woods, Vintage International
Edition, New York, 1997, p. 208
All quotes from John Cage in this section are taken from *Empty Words*: Writings
'73-'78, Marion Boyars, London, 1980
Norman O Brown, *Love's Body*, Random House, New York, 1966, p. 40

Invading Present Time
62.
William S Burroughs, 'Electronic Revolution', reprinted in *Ah Pook Is Here and
Other Texts*, John Calder, London, 1979, p. 130
David Gioavannoni, quoted in 'Oldest recorded voices sing again', BBC online
news report, last updated Friday, 28 March 2008, http://news.bbc.co.uk/1/
hi/7318180.stm, accessed 12.1.11
Jean Cocteau, *Two Screenplays: The Blood of the Poet, The Testament of
Orpheus*, translated by Carol Martin-Sperry, Marion Boyars, London and New
York, 1985, p. 73 and p. 83

63.
La Quotidienne, quoted in *Blood in the City*: Violence & Revelation in Paris
1789 – 1945, Richard D E Burton, Cornell University Press, New York, 2001, p. 278
Theodor W Adorno and Max Horkheimer, *Dialectic of Enlightenment*, translated
by John Cumming, Verso, London and New York, 1997, p. 11

Bruno Corradini, Emilio Settimelli, 'Weights, Measures and Prices of Artistic
    Genius – Futurist Manifesto 1914', *Futurist Manifestos*, edited by Umbro
    Apollonio, Thames & Hudson, London, 1973, p. 136
William S Burroughs, 'the invisible generation', reprinted in *The Job*: Topical
    Writings and Interviews, John Calder, London, 1984, p. 160

64.

Theodor W Adorno and Max Horkheimer, *Dialectic of Enlightenment*, translated
    by John Cumming, Verso, London and New York, 1997, p. 17
Jean Cocteau, *Two Screenplays: The Blood of the Poet, The Testament of
    Orpheus*, translated by Carol Martin-Sperry, Marion Boyars, London and New
    York, 1985, p. 77
Eric Norden, 'The Playboy Interview: Marshall McLuhan', *Playboy* Magazine,
    March 1969, http://www.nextnature.net/?p=1025, accessed online 28.10.09
William S Burroughs, 'the invisible generation', reprinted in *The Job*: Topical
    Writings and Interviews, John Calder, London, 1984, p. 163
William S Burroughs, extract chosen at random from *Nova Express*, Panther
    Science Fiction edition, 1968, p. 85
William S Burroughs, 'Interview with William S Burroughs', reprinted in William S
    Burroughs and Brion Gysin, *The Third Mind*, Viking Press, 1978, p. 1

65.

William S Burroughs, 'Precise Intersection Points', reprinted in William S
    Burroughs and Brion Gysin, *The Third Mind*, Viking Press, 1978, p. 135

66.

Theodor W Adorno and Max Horkheimer, *Dialectic of Enlightenment*, translated
    by John Cumming, Verso 1997, p. 68

67.

Unless otherwise identified, all quotes from William S Burroughs reproduced
in italics in this section are from 'the invisible generation', reprinted in *The Job*:
Topical Writings and Interviews, John Calder, 1984.
Thomas Alva Edison, quoted in Greg Milner, *Perfecting Sound Forever*: The Story
    of Recorded Music, Granta Books, London, 2009, p.34
William S Burroughs, 'Electronic Revolution', reprinted in *Ah Pook Is Here and*

Other Texts, John Calder, London, 1979, p. 129

68.
Theodor W Adorno and Max Horkheimer, *Dialectic of Enlightenment*, translated by John Cumming, Verso, London and New York, 1997, p. 84 and p. 121
Hans-Jürgen Syberberg, *Hitler*: A Film from Germany, translated by Joachim Neugroschel, Carcanet New Press Ltd, 1982, p.43
William S Burroughs, 'Electronic Revolution', reprinted in *Ah Pook Is Here and Other Texts*, John Calder, London, 1979, p. 126
William S Burroughs, 'It Belongs to the Cucumbers', *The Adding Machine*: Collected Essays, John Calder, London, 1985, p. 54
William S Burroughs, *Burroughs Live*: The Collected Interviews of William S Burroughs, Sylvère Lotringer, Semiotext(e), Los Angeles, 2001, p. 157

69.
Shinji Mikami, quoted in 'The Godfather', Zev Borow, *Wired*, 11.01 January 2003, p. 145

70.
Theodor W Adorno and Max Horkheimer, *Dialectic of Enlightenment*, translated by John Cumming, Verso, London and New York, 1997, pp. 126-7
Jeremy Relph, 'Broads, a Bitch, Never the Snitch', reproduced in *Game On*: The History and Culture of Videogames, edited by Lucien King, Laurence King, London, 2002, p. 58
Gilles Deleuze and Félix Guattari, *Anti-Oedipus*: Capitalism and Schizophrenia, translated by Robert Hurley, Mark Seem and Helen R Lane, The Viking Press, New York, 1977, p. 7

71.
Jared Lee Loughner, 'America: Your Last Memory In A Terrorist Country!', video posted on YouTube October 2, 2010, http://www.youtube.com/watch?v=IDiqo6K5ZA4, accessed 31.5.11
Carl Von Clausewitz, *On War*, edited with an introduction by Anatol Rapoport, Penguin Books, London, 1982, p. 116 and p. 121
Richard Wald, quoted in 'The Gray Lady of Cable News', Michael Wolff, *Vanity Fair* 601, September 2010, p. 92

William S Burroughs, 'the invisible generation', reprinted in *The Job*: Topical
    Writings and Interviews, John Calder, London, 1984, p. 162
William S Burroughs, 'Electronic Revolution', reprinted in *Ah Pook Is Here and
    Other Texts*, John Calder, London, 1979, p. 131

72.
Edgar Rice Burroughs, *Under the Moons of Mars*, University of Nebraska Press,
    Lincoln and London, 2003, p. 340
Jean Cocteau, *Two Screenplays: The Blood of the Poet, The Testament of
    Orpheus*, translated by Carol Martin-Sperry, Marion Boyars, London and New
    York, 1985, p. 74 and p. 89
'Coded American Civil War message in bottle deciphered', BBC News, US
    and Canada, 25 December 2010 http://www.bbc.co.uk/news/world-us-
    canada-12079281, accessed 12.2.11

73.
Nick Bostrom, 'Are You Living In A Computer Simulation?', *Philosophical
    Quarterly*, Vol. 53, No. 211, 2003, pp. 243-255 http://www.simulation-argument.
    com/simulation.html, accessed 7.6.11
Friedrich Nietzsche, *The Gay Science*, 341, translated by Josephine Nauckhoff,
    Cambridge Texts in the History of Philosophy, Cambridge University Press,
    Cambridge, 2001 by p. 194

Requiem for the Network
74.
'Iran's "Twitter revolution" was exaggerated, says editor,' Wednesday 9 June
    2010 18.40 BST http://www.guardian.co.uk/world/2010/jun/09/iran-twitter-
    revolution-protests accessed 7.7.11
W H Matthews, *Mazes and Labyrinths*: A General Account of Their History
    and Developments, Longmans in 1924, facsimile edition, The Lost Library,
    Glastonbury, undated, p. 94
Herbert Marcuse, *One-Dimensional Man*: Studies in the Ideology of Advanced
    Industrial Society, Routledge, Abingdon and New York, 2007, p. 55
Fred Vogelstein, 'Case Study: Microsoft', *Wired* 15.04, April 2007, p. 170

People's Daily, quoted in Evgeny Morozov, *The Net Delusion*: How Not to Liberate
the World, Allen Lane, London, 2011, p.12

75.

Herbert Marcuse, *One-Dimensional Man*: Studies in the Ideology of Advanced
Industrial Society, Routledge, Abingdon and New York, 2007, p. 20

Ada Lovelace, footnote to Louis Menabrea's *Sketch of an Analytical Engine
Invented by Charles Babbage*, quoted in Sadie Plant, *Zeros and Ones*:
Women, Cyberspace and the New Sexual Revolution, Fourth Estate, London,
1997, p. 11

Friedrich Nietzsche, letter toward the end of February 1882, quoted in
Gramophone, Film, Typewriter, Friedrich A Kittler, translated with an
Introduction by Geoffrey Winthrop-Young and Micheal Wutz, Stanford
University Press, California, 1999, p. 200

Carl Von Clausewitz, *On War*, edited with an introduction by Anatol Rapoport,
Penguin Books, London, 1982, pp. 164-5

Calottes et soutanes: Jésuites et Jésuitesses, quoted in *Blood in the City*:
Violence & Revelation in Paris 1789 – 1945, Richard D E Burton, Cornell
University Press, New York, 2001, p. 187

76.

John Cage, 'The Future of Music', *Empty Words*: Writings '73 – '78, Marion Boyars,
London, 1980, p. 183

Carl Von Clausewitz, *On War*, edited with an introduction by Anatol Rapoport,
Penguin Books, London, 1982, p. 199

The description of Le Corbusier's museum design taken from Jean-Louis Cohen,
'Le Corbusier's Nietzschean Metaphors', *Nietzsche and 'An Architecture of Our
Minds'*, edited by Alexandre Kostka and Irving Wohlfarth, The Getty Research
Institute Publications and Exhibitions Program, Los Angeles, 1999, p. 324

Herodotus, quoted in *Mazes and Labyrinths*: A General Account of Their History
and Developments, W H Matthews, Longmans in 1924, facsimile edition, The
Lost Library, Glastonbury, undated, p. 7

RANDom News, volume 9, number 1, quoted in *One-Dimensional Man*: Studies
in the Ideology of Advanced Industrial Society, Herbert Marcuse, Routledge,
2007, Abingdon and New York, pp. 84-5

77.
Carl Von Clausewitz, *On War*, edited with an introduction by Anatol Rapoport, Penguin Books, London, 1982, p. 227

78.
Paul Baran, in conversation with the author, 20.11.07
Jean-Luc Godard, *Alphaville*, translated by Peter Whitehead, Faber and Faber Ltd, London, 2000, pp. 43-44

79.
Jean-Luc Godard, *Alphaville*, translated by Peter Whitehead, Faber and Faber Ltd, London, 2000, p. 54
John Heider, 'Olden Times and New: Sex and Drugs at Esalen', unpublished essay, 1991, quoted in *Esalen*: America and the Religion of No Religion, Jeffrey J Kripal, The University of Chicago Press, Chicago and London, 2007, p. 14

80.
Carl Von Clausewitz, *On War*, edited with an introduction by Anatol Rapoport, Penguin Books, London, 1982, p. 242
Jean-Luc Godard, *Alphaville*, translated by Peter Whitehead, Faber and Faber Ltd, London, 2000, p. 62
Herman Khan and Anthony J Wiener, *The Year 2000*: A Framework for Speculation on the Next Thirty-Three Years, The MacMillan Company, 1967, p. 97 (Note: this quote has been redacted slightly for length)

81.
John Cage, 'Foreword', *X*: Writings '79 – '82, Wesleyan University Press, Connecticut, 1983, p. ix
Richard L Brandt, *Upside*, quoted by Tom Wolfe in his foreword to *Understanding Me*: Lectures and Interviews, Marshall McLuhan, MIT Press, Massachusetts, 2003, p. x

82.
W H Matthews, *Mazes and Labyrinths*: A General Account of Their History and Developments, Longmans in 1924, facsimile edition, The Lost Library, Glastonbury, undated, p. 111

83.

IM exchange between 'Sabu' and 'Virus', August 11, 2011, http://pastie.org/
private/om3mrqvbdbmg8esddkcmw, accessed 8.9.11

'Work-life Balance', *The Economist*, 23 December 2006-5 January 2007, p.113

Martin Libicki, quoted in 'Cyber Weapons: The New Arms Race' Michael Riley
and Ashlee Vance, *Business Week*, July 20, 2011, 11:45 PM EDT, http://www.
businessweek.com/magazine/cyber-weapons-the-new-arms-race-07212011.
html, accessed 23.9.11

Chris Cox, Facebook's VO of Product, quoted in Steven Levy, 'Exclusive: Inside
Facebook's Bid to Reinvent Music, News and Everything', *Wired* Epicenter,
September 22, 2011, http://www.wired.com/epicenter/2011/09/facebook-new-
profile-apps/, accessed 25.9.11

84.

W H Matthews, *Mazes and Labyrinths*: A General Account of Their History
and Developments, Longmans in 1924, facsimile edition, The Lost Library,
Glastonbury, undated, p. 26

Policy paper quoted in *The Net Delusion*: How Not to Liberate the World, Evgeny
Morozov, Allen Lane, London, 2011, p. 123

Theodor W Adorno and Max Horkheimer, *Dialectic of Enlightenment*, translated
by John Cumming, Verso, London and New York, 1997, p. 195

85.

IM exchange from Iraq between ZJ Antolak, aka 'Zinnia Jones' and Bradley
Manning aka 'bradass87', 21 February 2009, quoted in 'Bradley Manning's
Army of One', Xeni Jardin, posted on Boing Boing, http://boingboing.
net/2011/07/03/bradley-mannings-arm.html, accessed 3.7.11

Steven Fishman 'Bradley Manning's Army of One: How a lonely, five-foot-two,
gender-questioning soldier became a WikiLeaks hero, a traitor to the U.S., and
one of the most unusual revolutionaries in American history', *New York*, July
3, 2011 http://nymag.com/news/features/bradley-manning-2011-7/ accessed
3.7.11

'Work-life balance', *The Economist*, 23 December 2006-5 January 2007, p. 114

Gunter Ollmann and Dave Aitel, quoted in 'Cyber Weapons: The New Arms Race'
Michael Riley and Ashlee Vance, *Business Week*, July 20, 2011, 11:45 PM EDT,

http://www.businessweek.com/magazine/cyber-weapons-the-new-arms-race-07212011.html, accessed 23.9.11

Godzilla Has Left the Building
Note: this is not the first time I have written about Godzilla. For a general intro-
duction to my thinking on the subject, readers are directed to 'Tokyo Must Be
Destroyed', published in the anthology *Digital Delirium*, edited by Arthur and
Marilouise Kroker (MacMillan), 'Gojira, Mon Amour', originally published in *Sight
& Sound*, June 1998 and, for a more personal interpretation, 'Godzilla
Has Left the Building' in *Nude* Issue 7, Winter 2005.

86.
Sam Kashner, 'Elizabeth Taylor's Closing Act', *Vanity Fair* 610, June 2011, p. 108

87.
Chernobyl Tour Gide, accessed http://www.world66.com/europe/ukraine/
    chernobyl, 12.12.11
Günther Oettinger, quoted in Martin Evans and Gordon Rayner, 'Japan nuclear
    plant disaster: warning of an "apocalypse" as fallout hits danger levels', *Daily
    Telegraph*, 16 March 2011 http://www.telegraph.co.uk/news/worldnews/asia/
    japan/8384809/Japan-nuclear-plant-disaster-warning-of-an-apocalypse-as-
    fallout-hits-danger-levels.html, accessed 3.1.12
Mario Praz, *The Romantic Agony*, Second Edition, OUP, London and New York,
    1970, pp.21-22
W H Matthews, *Mazes and Labyrinths*: A General Account of Their History
    and Developments, Longmans in 1924, facsimile edition, The Lost Library,
    Glastonbury, undated, pp. 117 and 143
Takeno Suzuki, 'Parts of Sendai Destroyed But Not Its Humanity', Nichi Bei, 7 April
    2011, http://www.nichibei.org/2011/04/parts-of-sendai-destroyed-but-not-its-
    humanity/, accessed 27.11.11
Lautreamont's Maldoror, translated by Alexis Lykiard, Alison & Busby, London,
    1970, p. 30

88.
Susan Sontag, 'Imagination of Disaster', *Against Interpretation* and Other Essays,
    Dell, New York, 1966, pp. 215-216

Jacob Appelbaum, quoted in Heather Brooke, 'Inside the Secret World of the Hackers', *The Guardian*, 24 August 2011 http://www.guardian.co.uk/technology/2011/aug/24/inside-secret-world-of-hackers, accessed 9.11.11

89.

Alfred Jarry, quoted in Roger Shattuck, *The Banquet Years*, The Origins of the Avant-Garde in France 1885 to World War I, revised edition, New York, 1968, p. 240

Susan Sontag, 'Imagination of Disaster', *Against Interpretation and Other Essays*, Dell, New York, 1966, pp. 227

Kenneth Frampton, *Architectural Design* 35, February 1965, quoted in *Beyond Archigram*: The Structure of Circulation, Hadas A Steiner, Routledge, New York and Abingdon, 2009, pp. 92 and 113

90.

Jean-Luc Godard, *Alphaville*, translated by Peter Whitehead, Faber and Faber Ltd, London, 2000 p. 68

91.

Viktor Shklosvky, *Third Factory*, Introduction and Translation by Richard Sheldon, Afterword by Kyn Hejinian, Dalkey Archive Press, Illinois, 2002, p. 85

93.

Tim Lenoir and Henry Lowood, 'Theaters of War: The Military-Entertainment Complex' from Jan Lazardzig, Helmar Schramm, Ludger Schwarte (eds), *Kunstkammer, Laboratorium, Bühne-Schauplätze des Wissens im 17. Jahrhundert/ Collection, Laboratory, Theater*, Walter de Gruyter Publishers, 2003, online PDF printout p. 1 http://www.stanford.edu/class/sts145/Library/Lenoir-Lowood_TheatersOfWar.pdf, accessed 13.6.11

Jack Thorpe, in conversation with the author, 15.2.10

94.

Karlheinz Stockhausen, Hamburg press conference, 18 September 2001, quoted in Christian Hänggi, 'Stockhausen at Ground Zero', in *Fillip* 15, Fall 2011, http://fillip.ca/content/stockhausen-at-ground-zero, accessed 6.4.12

Karlheinz Stockhausen, 'Message from Professor Karlheinz Stockhausen',
     September 19, 2001, posted at http://www.stockhausen.org/message_from_
     karlheinz.html, accessed 6.4.12
Karlheinz Stockhausen, in conversation with the author, 12.3.1999
'Backup Tokyo: Japan looks at creating world's first backup city', editorial, World
     Architecture News, 31 October 2011 http://www.worldarchitecturenews.com/
     index.php?fuseaction=wanappln.projectview&upload_id=17908, accessed
     6.1.12

95.

Richard Wagner, libretto for *Götterdämmerung*, piano transcription by Otto
     Singer, Breitkopf & Härtel, 1914, reproduced in booklet notes for the Deutsche
     Grammaphon, issue 457 7906-2, p. 217
Jean-Luc Godard, *Alphaville*, translated by Peter Whitehead, Faber and Faber
     Ltd, London, 2000 p. 45
Maurice Blanchot, *The Writing of the Disaster*, translated by Anne Smock,
     University of Nebraska Press, 1986
Alfred Jarry, epigraph to *Ubu Enchainé, Tout Ubu*, Librarie Generale Francaise,
     Paris, 1962, p. 269, translation the author's own
John Ruskin, 'The Seven Lamps of Architecture', Selected Writings, Oxford World
     Classics, Oxford and New York, 2004, p. 25

Experimental Cognition

Note: the quotation from Nietzsche at the head of this chapter is presented
exactly as it appears in R J Hollingdale's *Nietzsche*: The Man and His Philosophy,
revised edition, Cambridge University Press, Cambridge, 1999, page 103. It has
been edited down from a much longer paragraph by the book's author, who is
also the text's translator. The complete statement can be found in Hollingsdale's
translation of Nietzsche's *Untimely Meditations* published in the Cambridge Texts
in the History of Philosophy series, Cambridge University Press, Cambridge, 1997,
across pages 149 and 150.

96.

All quotes from Edgar Allan Poe, 'The Man Who Was Used Up' are taken as they
     appear in *Selected Writings of Edgar Allan Poe*: Poems, Tales, Essays and

Reviews, edited with an introduction by David Galloway, Penguin Books, 1974, pp. 127-136

Karel Čapek, *R U R (Rossum's Universal Robots)*, reprinted in *Of Men and Machines*, edited by Arthur O Lewis Jr, Dutton Paperbacks, New York, 1963, p. 52

Peter Menzel and Faith D'Aluisio, *Robo sapiens*: Evolution of a New Species, The MIT Press, Massachusetts, 2000, p. 229

97.

E T A Hoffmann, 'The Sandman', from *Tales of E. T. A. Hoffmann*, edited and translated by Leonard J Kent and Elizabeth C Knight, The University of Chicago Press, Chicago and London, 1969, p. 117

Friedrich Nietzsche, *Untimely Meditations*, translated by R J Hollingdale, Cambridge Texts in the History of Philosophy series, Cambridge University Press, Cambridge, 1997, I/4, p. 19

Walter Benjamin, *The Arcades Project*, translated by Howard Liland and Kevin McKaughlin, The Belknap Press of Harvard University Press, Cambridge, Massachusetts and London, England, 1999, p. 559 (Note: Benjamin's term 'Lebenslüge', or 'life-sustaining lie', is in turn derived from his reading of Henrik Ibsen's *The Wild Duck*)

98.

Karel Čapek, English *Saturday Review* interview 1923, http://public.wsu.edu/~delahoyd/sf/r.u.r.html, accessed 29.6.12

Henry Ford, *My Life and Work*, Garden City: Doubleday, 1923, pp. 108-9, quoted in Darren Wershler-Henry, The Iron Whim: A Fragmented History of Typewriting, McLelland and Stewart, 2005, p. 143

All quotes taken from Comte de Villiers de l'Isle-Adam, *L'Eve Future* appear in the M. Brunhoff edition, Paris, 1886, accessed through Project Gutenberg, http://www.gutenberg.org/ebooks/26681 3.7.12, 23.6.12 (Note: author's own translation. Emphasis is as it appears in the original)

Thomas Edison, quoted in Gaby Wood, *Edison's Eve*: A Magical History of the Quest for Mechanical Life, Anchor Books, New York, 2003, p. 174

All quotes from Thea von Harbou, *Metropolis* are taken from the original 1927 English edition, reissued by Ace Books, New York, 1963, pp. 53-54

99.

Comte de Villiers de l'Isle-Adam, *L'Eve Future*, M. Brunhoff edition, Paris, 1886, accessed through Project Gutenberg, http://www.gutenberg.org/ebooks/26681, 3.7.12 (Note: author's own translation)

Marshall McLuhan, 'notes on the lectures of I A Richards', National Archives of Canada, volume 3, file 6, quoted in *Marshall McLuhan*: Escape Into Understanding, W Terrence Gordon, Basic Books, New York, 1997, p. 49

Marshall McLuhan, 'notes on the lectures of I A Richards', National Archives of Canada, no volume or file numbers given, quoted in *Marshall McLuhan*: The Medium and the Messenger, Philip Marchand, MIT Press, Massachusetts, 1998, p. 38

Except where otherwise indicated, quotes by Alan Turing in this section are taken from 'On Computable Numbers, with an Application to the *Entscheidungsproblem*', Proceedings of the London Mathematical Society, Series 2, 42 (1936–37), pp. 230–265

Jean Cocteau, *Two Screenplays: The Blood of the Poet, The Testament of Orpheus*, translated by Carol Martin-Sperry, Marion Boyars, London and New York, 1985, p. 73

100.

Jonathan Benthall, *Science and Technology in Art Today*, Thames and Hudson, London, 1972, p. 84

Alan Turing, 'Lecture on the Automatic Computing Engine', *The Essential Turing*, edited by Jack Copeland, Clarendon Press, Oxford, 2004, p. 393

101.

Ludwig Wittgenstein, *The Blue and Brown Books*: Preliminary Studies for the 'Philosophical Investigations', Blackwell Publishing, Oxford, 1969 p. 47

All quotes from Alan Turing in this section are taken from 'Computing machinery and intelligence', *Mind*, volume 59, no 236, October 1950, pp. 433-460, online transcript http://www.loebner.net/Prizef/TuringArticle.html, accessed 17.7.12

Thea von Harbou, *Metropolis*, original 1927 English edition, Ace Books, New York, 1963, Ace Books 1963, p. 53

102.

Exchange between 'Alan' and 'Sruthi', Creative Machines Lab, Cornell University, August 2011, transcript of Cleverbot conversation taken from Andrew Nusca, 'Turning Artificial Intelligence on Itself', Smart Planet, http://www.smartplanet.com/blog/smart-takes/turning-artificial-intelligence-on-itself-video/18846, accessed 16.7.12

'Theological Chatbots', interview with Jason Yosinski and Igor Labutov, The Technium, 31 August 2011, http://www.kk.org/thetechnium/archives/2011/08/theological_cha.php, accessed 21.7.12

See also: Mark Frauenfelder, http://boingboing.net/2011/08/31/short-interview-with-creators-of-cleverbot-avatar-video.html, accessed 16.7.12

Description of PARRY taken from *Where Wizards Stay Up Late*: The Origins of the Internet, Kaite Hafner and Matthew Lyon, Touchstone, New York, 1996, p. 182

Transcript of PARRY/DOCTOR, 'Parry Encounters the Doctor', Vincent Cerf, Network Working Group, 21 January 1973, http://tools.ietf.org/rfc/rfc439.txt, accessed 17.7.12

Jonathan Benthall, *Science and Technology in Art Today*, Thames and Hudson, London, 1972, p. 84

103.

'Morals and the Machine', *The Economist*, 2-8 June 2012, p. 13

All quotes from *Astro Boy* in this section taken from the English subtitles to Osamu Tezuka 1983 *Astro Boy* series, original Japanese version, released by Manga, 2005

Andy Warhol speaking to *Time* May 3, 1963, quoted in *Warhol*, Victor Bockris, Penguin Books, London, 1989, p.190

Andy Warhol, *The Andy Warhol Diaries*, edited by Pat Hackett, Warner Books, New York, 1989, p. 340

Alan Turing, 'Computing machinery and intelligence', *Mind*, volume 59, no 236, October 1950, p. 454

104.

Karel Čapek, *R U R (Rossum's Universal Robots)*, reprinted in *Of Men and Machines*, edited by Arthur O Lewis Jr, Dutton Paperbacks, New York, 1963, p. 38

Friedrich Nietzsche, *Ecce Homo*: How One Becomes What One Is, translated by R J Hollingdale, Penguin Books, London, 1992, p. 104

105.

Garry Kasparov, 'The Chess Master and the Computer' *The New York Review of Books*, February 11. 2010, http://www.nybooks.com/articles/archives/2010/feb/11/the-chess-master-and-the-computer/, accessed 12.6.12

Alan Turing, 'Lecture on the Automatic Computing Engine', *The Essential Turing*, edited by Jack Copeland, Clarendon Press, Oxford, 2004, p. 75

Quotes relating to DeepQA are taken from 'The DeepQA Project', IBM online statement, 27.4.11, http://www.research.ibm.com/deepqa/deepqa.shtml, accessed 24.6.12

Alan Turing, 'Computing machinery and intelligence', *Mind*, volume 59, no 236, October 1950, p. 433

106.

E T A Hoffmann, 'The Sandman', from *Tales of E. T. A. Hoffmann*, edited and translated by Leonard J Kent and Elizabeth C Knight, The University of Chicago Press, Chicago and London, 1969, p. 123

All quotes from Alan Turing in this section are taken from 'Computing machinery and intelligence', *Mind*, volume 59, no 236, October 1950, pp. 433-460.

All quotes from 'Alan' and 'Sruthi' taken from the transcript of Cleverbot conversation reproduced in Andrew Nusca, 'Turning Artificial Intelligence on Itself', *Smart Planet*, http://www.smartplanet.com/blog/smart-takes/turning-artificial-intelligence-on-itself-video/18846, accessed 16.7.12

'Oh, That's Near Enough', *The Economist*, 'Technology Quarterly', 2-8 June 2012, pp. 3-4

107.

Edgar Allan Poe, 'The Man Who Was Used Up', *Selected Writings of Edgar Allan Poe*: Poems, Tales, Essays and Reviews, edited with an introduction by David Galloway, Penguin Books, 1974, p. 130

'IBM Moves Closer To Creating Computer Based on Insights From The Brain', online press release, dated 18.11.09, http://www-03.ibm.com/press/us/en/pressrelease/28842.wss, accessed 24.6.12

Living in Space
108.

Wernher von Braun, *Congressional Record*, 10 October1963, quoted in Gerard

DeGroot, *Dark Side of the Moon*: The Magnificent Madness of the American Lunar Quest, Jonathan Cape, London, 2007, p. 145

110.
Tod Dockstader, in conversation with the author, 8.4.05
John Glenn, NASA press conference, April 9, 1959, quoted in Gerard DeGroot, *Dark Side of the Moon*: The Magnificent Madness of the American Lunar Quest, Jonathan Cape, London, 2007, p. 108
Manfred E Clynes and Nathan S Kline, 'Cyborgs and Space', *Astronautics*, September 1960, p. 27 PDF available from http://web.mit.edu/digitalapollo/Documents/Chapter1/cyborgs.pdf, accessed 29.8.12

111.
All quotes from Manfred E Clynes and Nathan S Kline, 'Cyborgs and Space', *Astronautics*, September 1960 taken from http://web.mit.edu/digitalapollo/Documents/Chapter1/cyborgs.pdf, accessed 10.9.12
Marcel Proust, *In Search of Lost Time* Volume IV: Sodom and Gomorrah, translated by C K Scott Moncrieff & Terence Kilmartin, edited by D. J. Enright, Vintage Books, London, 2000, p. 60
Erwin Schrödinger, *What Is Life?* The Physical Aspect of the Living Cell, Cambridge University Press, Cambridge, 1992, pp. 73 and 77

113.
Richard M Nixon, quoted in Charles Cooper 'The speech Nixon had prepped for an Apollo 11 disaster', CNet news August 27, 2012, http://news.cnet.com/8301-11386_3-57501134-76/the-speech-nixon-had-prepped-for-an-apollo-11-disaster/, accessed 27.8.12

114.
Washington Post, quoted in *The Economist*, 'The End of the Dream?', 16-22 August 2008, p. 39
F T Marinetti, 'Multiplied Man and the Reign of the Machine', from *Selected Writings*, edited with an Introduction by R W Flint, Secker & Warburg, London, 1972, p. 91

115.

Richard Lesher, *San Francisco Daily Examiner and Chronicle*, 17 March 1968, quoted in *Dark Side of the Moon*, Gerrard DeGroot, Jonathan Cape, London, 2007, p. 149

Rem Koolhaas, 'Almost Paradise', *Wired* 11.06, June 2003, p.152

T A Heppenheimer, What's To Do on a Saturday Night? From *Colonies in Space*, 1977, http://www.nss.org/settlement/ColoniesInSpace/colonies_chap11.html, accessed 15.10.12

116.

Kisho Kurakawa, *Each One A Hero*: The Philosophy of Symbiosis, Kodansha, 1997 http://www.kisho.co.jp/index.php, accessed 1.12.12

'New Ticks in our Biological Clock', Sputnik Observatory, 19 October 2009, http://blog.sptnk.org/2009/10/19/new-ticks-in-our-biological-clock/, accessed 29.11.12

'Walking Straight Into Circles', Sputnik Observatory, 25 September 2009, http://blog.sptnk.org/2009/09/25/walking-straight-into-circles/, accessed 14.10.12

117.

Malcolm C Smith *et al*, 'Apollo Food Technology', Section VI, Chapter 1, *Biomedical Results of Apollo*, NASA SP-368, 1975, http://lsda.jsc.nasa.gov/books/apollo/s6ch1.htm, accessed 27.12.12

118.

Ron Herron, 'A Walking City', *Achigram* 5, 'Metropolis', Autumn 1964, p. 17

Jonathan Benthall, *Science and Technology in Art Today*, Thames and Hudson, London, 1972, p. 80

Donald Michie and Rory Johnston, *The Creative Computer*: Machine Intelligence and Human Knowledge, Penguin Books 1984, PDF transcription http://www.senster.com/ihnatowicz/articles/creative_computer.pdf, accessed 18.12.12

'UWE scientists design first robot using mould', University of West England news release, 27 August 2009, http://info.uwe.ac.uk/news/UWENews/news.aspx?id=1553, accessed 27.6.12

119.

Bigelow Aerospace homepage, http://www.bigelowaerospace.com/
introduction.php, accessed 21.12.12

Astro Boy, 'Outer Spaceport R45', Episode #46, taken from the English subtitles
to Osamu Tezuka 1983 *Astro Boy* series, original Japanese version, released by
Manga, 2005

Robert T Bigelow, 'An Open Letter to NASA Administrator Charles Bolden From
Robert Bigelow' 14 October 2009 http://www.spaceref.com/news/viewnews.
html?id=1354, accessed 27.11.12

F T Marinetti, 'Multiplied Man and the Reign of the Machine', from *Selected
Writings*, edited with an Introduction by R W Flint, Secker & Warburg, London,
1972, p. 91

Mark Wigley, 'The Architectural Brain' in *Network Practices*: New Strategies in
Architecture and Design, edited by Anthony Burke and Therese Tierney,
Princeton Architectural Press, New York, 2007, p.36

Airlines for America homepage, http://www.airlines.org/Pages/Home.aspx,
accessed 24.12.12

'Living the Dream', *The Economist*, 1-7 May 2010, p. 55

Richard J Norton, 'Feral Cities', *Naval War College Review*, Autumn 2003,
p. 99 http://www.usnwc.edu/getattachment/9a5bddeb-e16e-48fc-b21a-
22515e79aaa9/Feral-Cities---Norton,-Richard-J-.aspx, accessed 27.9.12

120.

W H Matthews, *Mazes and Labyrinths*: A General Account of Their History
and Developments, Longmans in 1924, facsimile edition, The Lost Library,
Glastonbury, undated, p. 49

'March of the Robots', *The Economist*, 2-9 June 2-9, Technology Quarterly, p. 11

'Extraordinary Mind Vint Cerf Predicts the Future of Mobile', Sputnik Observatory
interview, 25 January 2010, http://blog.sptnk.org/2010/01/25/extraordinary-
mind-vint-cerf-predicts-the-future-of-mobile/, accessed 23.11.11

Marcel Proust, *In Search of Lost Time* Volume V: The Captive, translated by C K
Scott Moncrieff & Terence Kilmartin, edited by D J Enright, Vintage Books,
London, 2000, p. 291

The Uploaded Self: A Book of the Dead
Georg Christoph Lichtenberg, *The Waste Books*, translated with an introduction
    by R J Hollingdale, New York Review Books, New York, 1990, p. 37
Jason Rhodes and Paul Casciato, 'Europeans Bury "Digital DNA" Inside Mountain
    Stronghold', Reuters via *PC Mag*, 18 May 2010 http://www.pcmag.com/
    article2/0,2817,2363904,00.asp, accessed 11.12.12
Ed Young, 'Synthetic double-helix faithfully stores Shakespeare's sonnets', *Nature*,
    23 January 2013, http://www.nature.com/news/synthetic-double-helix-
    faithfully-stores-shakespeare-s-sonnets-1.12279, accessed 4.6.13
Extract from President Carter's Voyager Golden Disc text quoted in 'Howdy
    Stranger', Jet Propulsion Laboratory, California Institute of Technology website,
    19 August 2002, http://www.jpl.nasa.gov/news/news.php?feature=555,
    accessed 13.5.13
Norman O Brown, *Love's Body*, Random House, New York, 1966, p. 38
'Loughner Sentenced to Life in Prison Without Parole', CNN News Blog, 8
    November 2012, 04:31 PM ET http://news.blogs.cnn.com/2012/11/08/victims-
    enter-court-for-giffords-shooter-sentencing/, accessed 13.5.13
Erwin Schrödinger, *What Is Life?* The Physical Aspect of the Living Cell,
    Cambridge University Press, Cambridge, 1992, p. 88
Friedrich Nietzsche, *The Gay Science*, 341, translated by Josephine Nauckhoff,
    Cambridge Texts in the History of Philosophy, Cambridge University Press,
    Cambridge, 2001, p. 194
Georg Christoph Lichtenberg, *The Waste Books*, p. 164

Index

© Strange Attractor Press 2014